Methods in Molecular Biology

For further volumes:
http://www.springer.com/series/7651

Animal Endo-siRNAs

Methods and Protocols

Edited by

Andreas Werner

RNA Biology Group, Institute for Cell and Molecular Biosciences, Newcastle University, Newcastle, UK

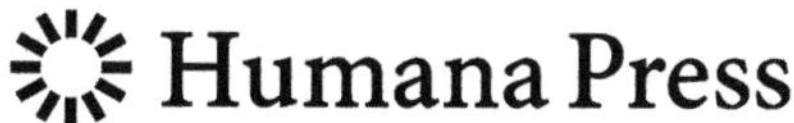

Editor
Andreas Werner
RNA Biology Group, Institute for Cell and Molecular Biosciences
Newcastle University, Newcastle, UK

ISSN 1064-3745 ISSN 1940-6029 (electronic)
ISBN 978-1-4939-0930-8 ISBN 978-1-4939-0931-5 (eBook)
DOI 10.1007/978-1-4939-0931-5
Springer New York Heidelberg Dordrecht London

Library of Congress Control Number: 2014941237

Printed on acid-free paper

Humana Press is a brand of Springer
Springer is part of Springer Science+Business Media (www.springer.com)

Preface

Short interfering RNAs (siRNAs) are widely used in biomedical research to modulate gene expression in a wide spectrum of cell types. The fact that *exogenous* siRNAs elicit a predictable effect when transfected into a cell, usually the knock down of a specific target mRNA, suggests that endogenous pathways utilizing *endogenous* siRNAs are in place. These short RNA oligonucleotides, now called endo-siRNAs, however, proved a rather elusive species in mammalian cells. Only the recently developed RNAseq technologies provided sufficient sequencing depth to comprehensively demonstrate and map endo-siRNAs in vertebrate animal systems. Yet, their biological role is still controversial. In non-vertebrates such as *Caenorhabditis elegans* and the fruit fly *Drosophila melanogaster* the molecular basics of endo-siRNAs are well established, both with regard to their molecular mechanism of synthesis and their biological function(s). As a consequence, researchers from separate research communities may define the term "endo-siRNA" in different ways.

The generally low expression level of endo-siRNAs in specific tissues and their close resemblance to breakdown products of endogenous cellular RNAs represent formidable challenges in this field. Moreover, detection of endo-siRNAs often involves reverse transcription, a step that produces low levels of false positive signals when strand specificity is of experimental importance—which is the case in studies focusing on endo-siRNAs. These difficulties emphasize the importance of established highly sensitive protocols to monitor small RNAs, specifically strategies for high-throughput sequencing, array technologies, and northern blotting (*see* Chapters 8–11 in this book). A further important consideration in the research of endo-siRNAs is the choice of model systems. As indicated above, *C. elegans* is by far the best established model system to study endo-siRNAs: Here, they have distinct structural features (length and defined first base, i.e., 26G siRNAs) and their regulatory role in gene expression is well documented (*see* Chapters 6 and 7 in this book). In the fruit fly *Drosophila* the synthesis of endo-siRNAs involves a specific Dicer molecule (Dicer2) and the pathway is essential to ensure normal fly development under perturbed environmental conditions (*see* Chapters 4 and 5 in this book). However, in vertebrates endo-siRNAs have only recently been detected and their biological role is unclear.

The aim of this book is to present a variety of approaches to investigate endo-siRNAs. Many of the protocols are generally applicable to study short RNAs expressed at a low level. Others highlight model systems that are particularly suitable to investigate specific aspects of endo-siRNAs, their synthesis, their genomics, or regulatory role. Unfortunately, there are no "one fits for all" methods when dealing with endo-siRNAs, especially in vertebrates. Consequently, this book suggests a compilation of applicable strategies for researchers entering the field of endo-siRNAs. Importantly, the book also contains many practical tips that are absent in standard lab manuals. In that sense, I hope the different contributions stimulate research and progress in the field of endo-siRNAs.

To conclude, I would like to sincerely thank all the contributors who shared their knowledge and experience and took their time and effort to realize the book on endo-siRNAs. Moreover, special thanks go to all the helpful individuals who assisted the preparation of this book.

Newcastle, UK *Andreas Werner*

Contents

Contributors

SAMMER ALNUMEIR • *RNA Biology Group, Institute for Cell and Molecular Biosciences, Newcastle University, Newcastle, UK*
CHRISTOPHE ANTONIEWSKI • *Laboratoire de Biologie du Développement, Drosophila Genetics and Epigenetics, CNRS UMR 7622 and University Paris 6 Pierre et Marie Curie, Paris, France*
MARIUS VAN DEN BEEK • *Drosophila Genetics and Epigenetics, Université Pierre et Marie Curie, Paris, France*
MARK CARLILE • *Faculty of Applied Sciences, University of Sunderland, Sunderland, UK*
CLÉMENT CARRÉ • *Drosophila Genetics and Epigenetics, Université Pierre et Marie Curie, Paris, France*
LONG CHEN • *John Curtin School of Medical Research, The Australian National University, Canberra, Australia*
JANE E. DAHLSTROM • *John Curtin School of Medical Research, The Australian National University, Canberra, Australia*
CAIO DAMSKI • *Biotechnology and Biomedical Sciences, The University of New South Wales, Sydney, Australia*
MARTIN DUFVA • *DTU Nanotech, Technical University of Denmark, Lyngby, Denmark*
KATHARINA ELMER • *Gene Center and Department of Biochemistry, Ludwig-Maximilians-Universität München, Munich, Germany*
KLAUS FÖRSTEMANN • *Gene Center and Department of Biochemistry, Ludwig-Maximilians-Universität München, Munich, Germany*
MONIKA GULLEROVA • *Sir William Dunn School of Pathology, University of Oxford, Oxford, UK*
SARAH E. HALL • *Department of Biology, Syracuse University, Syracuse, NY, USA*
SUNG-JUN HAN • *Drug Biology Group, Institut Pasteur Korea, Seongnam-Si, Gyeonggi-Do, South Korea*
STEPHANIE HELFER • *Gene Center and Department of Biochemistry, Ludwig-Maximilians-Universität München, Munich, Germany*
KIM HOLMSTRØM • *Bioneer A/S, Hørsholm, Denmark*
DAVID T. HUMPHREYS • *Molecular, Structural and Computational Biology Division, Victor Chang Cardiac Research Institute, Darlinghurst, Australia*
CRAIG P. HUNTER • *Department of Molecular and Cellular Biology, Harvard University, Cambridge, MA, USA*
GYORGY HUTVAGNER • *Faculty of Engineering and Information Technology, Centre of Health Technologies, University of Technology, Sydney, Australia*
JINGMIN JIN • *Division of RNA Biology, New England Biolabs, Ipswich, MA, USA*
BINO JOHN • *Department of Computational and Systems Biology, University of Pittsburgh School of Medicine, Pittsburgh, PA, USA*
SIMON P. KEAM • *Faculty of Engineering and Information Technology, Centre of Health Technologies, University of Technology, Sydney, Australia*

NOORA KOTAJA • *Department of Physiology, Institute of Biomedicine, University of Turku, Turku, Finland*
NELSON C. LAU • *Department of Biology, Rosenstiel Center of Basic Biomedical Sciences, Brandeis University, Waltham, MA, USA*
YOUNG SIK LEE • *College of Life Sciences and Biotechnology, Korea University, Seoul, South Korea*
ZHIHUA LI • *Department of Computational and Systems Biology, University of Pittsburgh School of Medicine, Pittsburgh, PA, USA*
DO-HWAN LIM • *College of Life Sciences and Biotechnology, Korea University, Seoul, South Korea*
TERESA T. LIU • *Department of Computational and Systems Biology, University of Pittsburgh School of Medicine, Pittsburgh, PA, USA*
LARRY A. MCREYNOLDS • *Division of RNA Biology, New England Biolabs, Ipswich, MA, USA*
OLIVER MEIKAR • *Department of Physiology, Institute of Biomedicine, University of Turku, Turku, Finland*
MILIJANA MIRKOVIC-HÖSLE • *Gene Center and Department of Biochemistry, Ludwig-Maximilians-Universität München, Munich, Germany*
KEVIN V. MORRIS • *Biotechnology and Biomedical Sciences, The University of New South Wales, Sydney, NSW, Australia; Molecular and Experimental Medicine, The Scripps Research Institute, La Jolla, CA, USA*
CHUN-TAEK OH • *Drug Biology Group, Institut Pasteur Korea, Seongnam-Si, Gyeonggi-Do, South Korea*
MARIA C. OW • *Department of Biology, Syracuse University, Syracuse, NY, USA*
DANNY RANGASAMY • *John Curtin School of Medical Research, The Australian National University, Canberra, Australia*
PHILIP K. SHIU • *Department of Molecular and Cellular Biology, Harvard University, Cambridge, MA, USA*
ANDREW SOBALA • *Department of Infectious Diseases, King's College London School of Medicine, London, UK*
MARTIN JENSEN SØE • *Bioneer A/S, Hørsholm, Denmark*
CATHERINE M. SUTER • *Molecular, Structural and Computational Biology Division, Victor Chang Cardiac Research Institute, Darlinghurst, Australia*
ANDREAS WERNER • *RNA Biology Group, Institute for Cell and Molecular Biosciences, Newcastle University, Newcastle, UK*
JIMMY J. ZHUANG • *Department of Molecular and Cellular Biology, Harvard University, Cambridge, MA, USA*

Chapter 1

Targeted Small Noncoding RNA-Directed Gene Activation in Human Cells

Caio Damski and Kevin V. Morris

Abstract

A growing body of evidence suggests that noncoding RNA (ncRNA) transcripts play a fundamental role in regulating gene expression via targeting epigenetic modifications to particular loci in the genome. Classical examples of such regulation are X-chromosome inactivation and genomic imprinting; however it is now clear that ncRNAs exert their influence over a wider array of genes throughout the metazoan genome. Accumulating evidence suggests that the ncRNAs act as guides for epigenetic silencing complexes to specific sites within the genome. Those ncRNAs involved in regulating the expression of particular protein-coding genes offer panoply of targets that when suppressed can result in derepression or activation of the ncRNA-targeted locus. Recent work has determined the underlying mechanisms involved in ncRNA-targeted epigenetic regulation in a subset of genes. These findings have resulted in a paradigm shift whereby targeted gene activation can be achieved, by targeting endogenous regulatory ncRNAs, producing potential novel treatments for genetic and infectious diseases where increases in gene expression are required.

Key words Long noncoding RNA, Antisense transcripts, Small noncoding RNAs, X chromosome inactivation, Genomic imprinting, Gene silencing, Gene activation, DNA methylation, Histone modification

1 Introduction

Over the last several years, evidence has steadily accumulated that RNA, once thought to be a mere intermediate between DNA and proteins in the "central dogma of molecular biology," in fact plays a fundamental role in coordinating epigenetic regulation in eukaryotes [1]. Estimates suggest that there are about four times as many noncoding RNA (ncRNA) transcripts as there are protein-coding RNA transcripts in the human cell, indicating an underappreciated role and possibly a functional role for these transcripts, as evolution would have favored against such gratuitous and wasteful transcription [2]. The functionality of ncRNA was further bolstered by the discovery that small noncoding RNAs (sncRNAs) have been found to be capable of mediating both transcriptional gene silencing (TGS)

Andreas Werner (ed.), *Animal Endo-siRNAs: Methods and Protocols*, Methods in Molecular Biology, vol. 1173, DOI 10.1007/978-1-4939-0931-5_1, © Springer Science+Business Media New York 2014

and activation in human cells [3]. The role of these ncRNA sequences has manifested themselves in a variety of processes, classic examples being X-chromosome inactivation (XCI) in female mammals and genomic imprinting. However regulation by long and short noncoding RNAs is not confined to a small, albeit interesting, subset of genes, but rather may represent a major contributor of gene regulation throughout an organism's growth, development, and perhaps even adaptation to a changing environment [4–6].

2 Long ncRNAs

The idea that RNA is involved in regulating gene expression has been around for almost five decades [7, 8]. Now whole genome/transcriptome sequencing, and collaborative research projects such as ENCODE, have begun to reveal just how pervasive such transcription and presumably regulation may be in human cells. An emerging class of regulatory RNAs are the long noncoding RNA (lncRNA); defined, somewhat arbitrarily, as nonprotein-coding RNA transcripts >200 nt in length [9]. Although the cut-off length has no empirical basis, it is nonetheless helpful in differentiating lncRNA from their shorter ncRNA cousins such as short-interfering RNAs (siRNAs), microRNAs (miRNAs), and Piwi-interacting RNAs (piRNAs). Recent analysis of the human transcriptome, such as the FANTOM3 project, has also shown that lncRNAs not only are abundant [10], but also exhibit a remarkable degree of tissue-specific transcription, particularly during development [11].

3 X-Chromosome Inactivation

XCI is a dosage compensation mechanism seen in female mammals, whereby an entire X-chromosome is rendered inoperative, thereby reducing the expression of X-linked genes to that seen in their male counterparts [12–15]. It is now apparent that this remarkable feat of transcriptionally silencing some ~1,000 genes is coordinated in *cis* by a single 100–500 kb region rich in lncRNA, termed the X-inactivation center (Xic) [16–18].

XCI requires that two chromosomes, sharing a common cell nucleus, follow diametrically opposed paths during development. Accurately coordinating this task is far from straightforward and presents three main challenges that must be resolved for this mechanism to be viable. The first is "counting"; determining the number of X-chromosomes present and initiating XCI if and only if there is more than one X residing in the nucleus [19–22]. The second is the problem of "choice"; the decision regarding which chromosome to silence necessitates some mechanism which operate in *trans* between the two X-chromosomes, to ensure that the decision is mutually exclusive, and avoid the lethal XaXa or XiXi genetic state. Lastly, the silencing factors must be localized to the chosen X and operate in a *cis*-dependent manner along the entirety of the X-chromosome while ignoring the other X [23].

4 ncRNA Regulators of XCI

Xist is a 17 kb lncRNA, which is largely responsible for initiating the compaction of heterochromatin and thus XCI. It appears that the accumulation of Xist around the chosen Xi creates a support structure and recruits silencing complexes in *cis*, causing the widespread changes to the chromatin along the X-chromosome. The major protein complex involved in this process is the Polycomb Repressive Complex (PRC2). A transcriptionally distinct 1.6 kb ncRNA known as Repeat A (RepA) is located at the 5′ end of the Xist locus. RepA contains two stem-loop structures as a 7.5 tandem repeat and binds Polycomb proteins, which are then directed to the Xic, where it promotes the expression of Xist via Histone 3 lysine 27 tri-methylation (H3K27me3) [14].

Xist and RepA are themselves subjected to regulation by another lncRNA, Tsix; a 40 kb lncRNA found antisense and overlapping to Xist. In undifferentiated cells, both Xist and Tsix are co-expressed from both X-chromosomes. At this stage, Tsix expression is 10- to100-fold higher than Xist expression; however during cell differentiation a choice is made for one chromosome to stop expressing Tsix. This chromosome is then destined to become Xi, while expression continues from the other chromosome and remains active. Without Tsix, Xist expression becomes upregulated and XCI can begin. This decision to stop or continue expression after differentiation is determined by Xite, located upstream of Tsix. Xite appears to act as an enhancer of Tsix expression, possibly involving a transcriptional-based mechanism. Thus, Xist and RepA promote XCI, while Xite and Tsix suppress it in a tug-of-war to decide the fate of the X-chromosome. Recent studies suggest that for a brief period after cell differentiation has begun, but before any increase in Xist expression, the two X-chromosomes pair-up at the Xic region [24]. This pairing may enable communication between the X-chromosomes [24–26]. It has been shown that a decrease in OCT4 can initiate XCI; adding to this, there is evidence to suggest that "counting" is regulated by an interaction between OCT4 and Xite. As this region is also implicated in "choice," it has been suggested that the pairing mechanism would assist intra-chromosomal communication, ensuring that the decision for Xi and Xa is mutually exclusive. In experiments where pairing is impeded, the cells become either XaXa or XiXi and are no longer viable.

5 Genomic Imprinting

Genomic imprinting is an epigenetic phenomena, involving histone modification or DNA methylation, which results in TGS on either the maternal or paternal chromosome, resulting ultimately in mono-allelic gene expression [27]. Imprinted genes tend to occur in clusters, although it is not infrequent to find non-imprinted genes residing within these domains [28]. Imprinting clusters are usually regulated by a *cis*-acting imprinting control region (ICR), which controls the behaviors of each gene within that cluster.

So far, approximately 100 imprinted genes have been identified in mammals [29], although it has been speculated that as much as 50 % of genes may be regulated by some form of imprinting. Imprinting plays an important role in regulating gene expression, particularly during growth and development. This has many implications for our understanding of diseases in both humans and animals. Because of their differential expression patterns, mutations affecting imprinted genes can present with different pathologies depending on whether or not the mutation was inherited from the maternal or paternal allele. There are a number of remarkable similarities between genomic imprinting and XCI [30]. Both share a common theme of being rich in ncRNA and have gene clusters, which are co-regulated, by a common *cis*-acting "switch." Both also show interactions with the Polycomb group of proteins (PcG), as well as the insulator CTCF.

6 Igf2 and H19 (Imprinted Loci)

The H19/Igf2 domain is a particularly well-characterized imprinted locus. Ifg2 and H19 are transcriptionally mutually exclusive, that is, Ifg2 is expressed exclusively from the paternal chromosome, while H19 is always expressed from the maternal chromosome. Igf2 (insulin-like growth factor 2) is a major growth factor in fetal development and H19 is an lncRNA transcript. Despite H19 being highly expressed in embryos and showing a high degree of conservation among mammals, no obvious phenotype is associated with its deletion. However, it has been proposed that the H19 may act as a miRNA precursor [31].

7 Mechanism of Antisense ncRNA-Directed Epigenetic Regulation

Understanding the fundamental mechanisms that underlie ncRNA and targeted gene suppression potentially offers a radical and novel way of treating a variety of diseases. Using our understanding of how long antisense ncRNAs cause epigenetic modifications to transcriptionally silence specific genes, it should be possible to

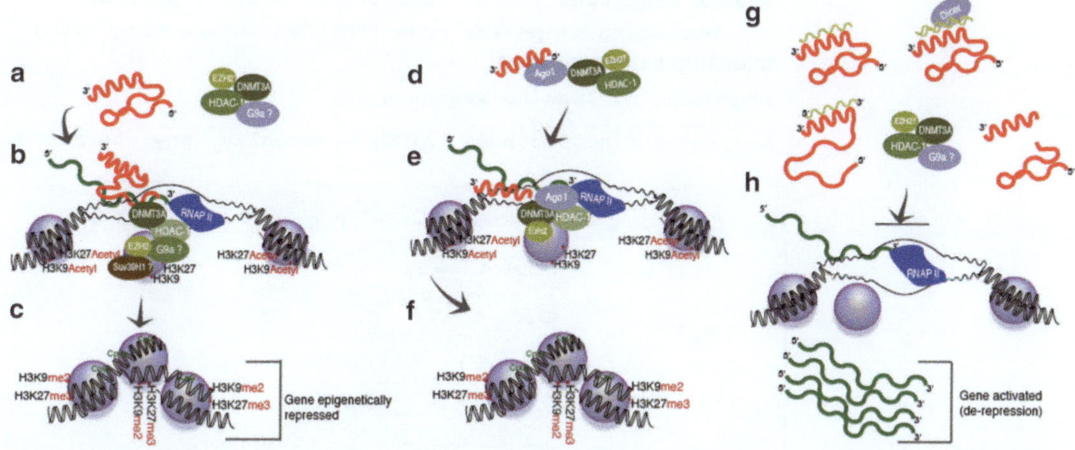

Fig. 1 *Mechanism of ncRNA-directed epigenetic regulation of gene expression.* (**a–c**) The endogenous lncRNA pathway of epigenetic regulation. (**a**) The lncRNA interacts with epigenetic remodeling proteins such as DNMT3a [44] and possibly other DNMT3a-associated proteins such as HDAC1 and EZH2 (reviewed in ref. 35). (**b**) The lncRNA/protein complex is then localized to the homology containing target loci. (**c**) The complex results in histone and DNA methylation at the targeted loci and ultimately chromatin compaction and transcriptional silencing. (**d–f**) Small antisense ncRNAs can be generated to mimic lncRNAs and drive gene silencing. (**g–h**) Suppression of endogenous lncRNAs by (**g**) antisense oligonucleotides or siRNAs can result in a loss of lncRNA function and (**h**) derepression of the lncRNA-targeted locus, ultimately resulting in activation of gene expression

design RNA sequences to mimic this endogenous mechanism for therapeutic purposes. Such therapies could be used to either suppress or enhance gene expression at specific loci without changes to the underlying genetic code. Targeted gene silencing would mimic the endogenous lncRNA pathway using small-antisense RNAs targeted to the promoters of specific genes to induce modifications to the chromatin such that gene expression is inhibited (Fig. 1a–c). Genes that are regulated endogenously by antisense lncRNA can also be targeted with small-interfering RNA (siRNA) or antisense oligonucleotides to degrade the target lncRNA thereby derepressing the gene and increase transcription levels [32, 33] (Fig. 1g–h).

8 Small ncRNA-Targeted Gene Silencing: Mimicking the lncRNA-Targeting Pathway to Epigenetically Silence a Genes Expression

The majority of studies in human cells using small ncRNAs to target gene expression have focused on siRNA- or miRNA-mediated posttranscriptional gene silencing, where short nucleotide sequences in combination with an Argonaute protein are used to target the 3′-end of mRNAs in a homology-dependent manner. This small ncRNA–protein complex prevents translation or induces

```
Search sequences for strings of consecutive purines.
    beginning of purine code (RNA:RNA interacting site)
NNNNRRRRRRRNNNNNNNNNN
beginning of leading sequence
R = any purine (AG)
N = any nucleotide base (ATCG) - actually, any character.
```

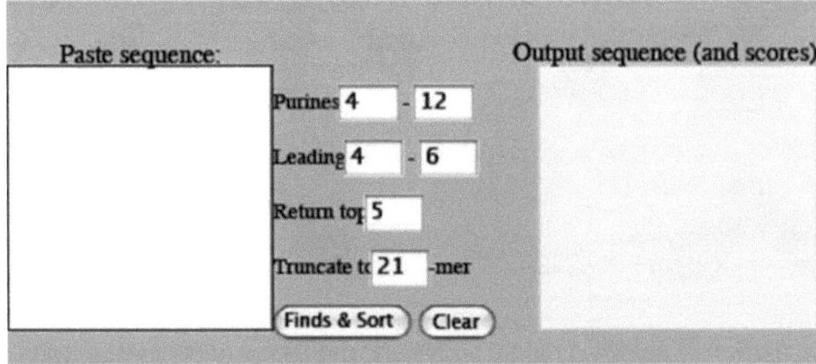

Fig. 2 Algorithm for targeted small antisense RNA-directed gene silencing. The above algorithm can be used to determine sasRNA target sites with a high likelihood of directing transcriptional and epigenetic silencing [36]. Access to this algorithm is available at: (http://www.morrislab.unsw.edu.au)

enzymatic cleavage of the targeted transcripts, in effect inhibiting gene expression [34]. However, there is also evidence that some aspects of this mechanism can be appropriated for inducting TGS via targeted epigenetic modifications at a specific locus [35].

Recently it has been shown that small antisense RNAs can be designed to mimic endogenous lncRNAs using a computer algorithm [36] (Fig. 2). This study found that both short antisense RNAs, and natural lncRNAs, which effectively induce TGS, often contain strings of purine residues 4–12 bp long. By hypothesizing that this may contribute to the effectiveness of TGS a variety of short antisense RNAs were tested on four selected genes. Sequences designed based on their purine content were then subjected to mutational comparison, changing the purine content as well as predicted structural folds of the small ncRNA. Sequences designed using this algorithm (Fig. 2) resulted in effective TGS in over 50 % of test cases. Preliminary results suggest that changes to the purine sequences generally reduce TGS, and that certain sequences are favored in stem-loop structures; however further studies are necessary to validate these observations. While still in the development and testing stages, such algorithms offer a promising means of designing effective small ncRNAs for targeted TGS.

Cases such as Tsix regulating Xist seen in Xci or genomic imprinting poignantly demonstrate the capacity for lncRNA to direct site-specific chromosomal changes that affect gene expression. Exploring the possibility of using small antisense RNAs to mimic these endogenous lncRNAs, a study by Hawkins et al. showed that continuously targeting the human ubiquitin C (UbC) gene promoter for 3 days leads to an increase in histone three lysine 27 trimethylation (H3K27me3) and DNA methylation, resulting in stable gene silencing at the targeted locus [37]. This study also corroborated findings in previous studies, demonstrating the necessity of Argonaute 1 (Ago1) [38], Histone Deacetylase 1 (HDAC-1) [39], DNA methyltransferase 3a (DNMT3a) [40], and a promoter-associated RNA [41] in effecting small antisense RNA-directed specific histone methylation and gene silencing (reviewed in ref. 42) (Fig. 1d–f). It was also elucidated in this study that histone methylation and Ago1 were triggered by the initiation of silencing whereas DNMT3a and DNA methylation were required for maintaining the silent state.

Although studies regarding TGS in human cells lag behind those of *Schizosaccharomyces pombe* or *Arabidopsis*, there exist many parallels between these well-characterized organisms, suggesting a common mode of action involving small ncRNAs and chromatin-modifying proteins; it is therefore conceivable to hypothesize similar mechanisms in human cells [43]. It has been shown that in human cells TGS is dependent on methylation of H3K9, H3K27, and DNA methylation at the targeted gene promoters (reviewed in ref. 42). DNMT3a is believed to be primarily responsible for establishing this local DNA methylation; by binding to small ncRNAs directly [40], DNMT3a can be recruited to and target the ncRNA homology-containing loci and establish methylation [40]. More recently DNMT3a has been found to bind directly to the PTENpg1 antisense lncRNA [44]. Interestingly, DNA methylation correlates with long-term silent chromatin states and is also found to be heritable or passed on to daughter cells [43]. Collectively, these observations suggest that ncRNAs and DNMT3a may be involved in the recently observed transgenerational inheritance, possibly acting as drivers of the acquirement of epigenetic states which are passed on to daughter cells [45].

9 Small ncRNA-Targeted Gene Activation: Targeting Antisense lncRNAs with Antisense Oligos or RNAi to Activate Gene Expression

It is not always the case however that promoter-targeted small ncRNAs induce TGS; in some instances siRNAs have been observed to induce transcriptional gene activation (TGA). One observation noted that when promoters, which had a low GC content and high complexity, were targeted with siRNAs, gene activation occurred [38, 46]. In this case, the activation appeared to be the result of changes to the "histone code," with at least one example of stable long-term activation [47], though the mechanism of action remains unknown.

In a set of studies built on these first observations of siRNA-directed TGS it was learned that the observed activation was the result of bidirectional transcription at the siRNA target locus. Interestingly, a long-ncRNA antisense to the protein-coding gene appeared to be required for the siRNA-directed TGA [48]. The activating siRNAs were designed to target this long antisense ncRNA, thereby resulting in the derepression of gene expression (Fig. 1g–h). This observation suggested that by inhibiting endogenous lncRNA gene repressors, gene expression is effectively increased [48, 49]. TGA using this mechanism has been demonstrated in studies focusing on p21, p15, and the progesterone receptor [46, 48, 50, 51], Oct4 [51], and most recently with PTEN, which was found to be regulated by an antisense lncRNA emanating from the PTEN pseudogene [44] (Fig. 1g–h).

A comparable mechanism of gene activation via derepression has also been demonstrated by Modarresi et al. In their study they observed increased levels of the brain-derived neurotropic factor (BDNF) mRNA and protein after targeted degradation of the BDNF-AS transcript, both in vitro and in vivo [52]. Using transfected siRNA to induce derepression, they were able to increase BDNF transcript levels by two- to sevenfold. Interestingly, the same principal was applicable to other genes such as glial-derived neurotropic factors (GDNF) and ephrin receptor B2 (EPHB2). Targeted suppression of the respective endogenous antisense transcripts leads to increased gene expression, accompanied by changes to the chromatin structure, more specifically, a significant reduction in H3K27me3 at the BDNF locus. Such derepression and activation of gene expression has also been observed with sasRNAs targeted to promoters of endogenous regulatory lncRNAs [36]. In this case, short antisense RNAs generated according to the algorithm introduced above (Fig. 2) targeted the promoter region for the lncRNA involved in regulating DUSP6, a tumor suppressor gene that is epigenetically silenced in pancreatic cancers [36].

These studies provide the proof of principle that siRNAs are able to induce changes to the chromatin structure, and either silence or upregulate specific targeted gene expression in a stable and long-term manner. These observations not only shed light on this emerging phenomenon of RNA-mediated mechanisms in regulating gene expression, but may also offer novel means of treating a variety of diseases. The benefits of such therapeutic options are obvious; generating short RNA oligonucleotides is cheap, easy, and rapid, and also offers the prospect of personalized medicine: Turning genes on or off in an epigenetic manner according to the patients-specific genetic footprint [32] (Fig. 1).

Acknowledgements

The project was supported by NIHLB R01AI084406, NIAID R56 AI096861-01, and PO1 AI099783-01 to KVM.

References

1. Lee J (2012) Epigenetic regulation by long noncoding RNAs. Science 338:1435–1439
2. Ponting C (2008) The functional repertoires of metazoan genomes. Nat Rev Genet 9:689–698
3. Morris K (2009) RNA-directed transcriptional gene silencing and activation in human cells. Oligonucleotides 19:299–306
4. Kung J, Colognori D, Lee J (2013) Long noncoding RNAs: past, present, and future. Genetics 193:651–669
5. Morris K (2009) Non-coding RNAs, epigenetic memory and the passage of information to progeny. RNA Biol 6(3):242–247
6. Rassoulzadegan M, Grandjean V, Gounon P, Vincent S, Gillot I, Cuzin F (2006) RNA-mediated non-mendelian inheritance of an epigenetic change in the mouse. Nature 441: 469–474
7. Davidson B, Fasman GD (1969) The single-stranded polyadenylic acid-poly-L-lysine complex. A conformational study and characterization. Biochemistry 8:4116–4126
8. Jacob F, Monod J (1961) Genetic regulatory mechanisms in the synthesis of proteins. J Mol Biol 3:318–356
9. Banfai B, Jia H, Khatun J, Wood E, Risk B, Gundling WE Jr, Kundaje A, Gunawardena HP, Yu Y, Xie L, Krajewski K, Strahl BD, Chen X, Bickel P, Giddings MC, Brown JB, Lipovich L (2012) Long noncoding RNAs are rarely translated in two human cell lines. Genome Res 22:1646–1657
10. Carninci P, Kasukawa T, Katayama S, Gough J, Frith M, Maeda N, Oyama R, Ravasi T, Lenhard B, Wells C, Kodzius R, Shimokawa K, Bajic V, Brenner S, Batalov S, Forrest AR, Zavolan M, Davis M, Wilming L, Aidinis V, Allen J, Ambesi-Impiombato A, Apweiler R, Aturaliya R, Bailey T, Bansal M, Baxter L, Beisel K, Bersano T, Bono H, Chalk A, Chiu K, Choudhary V, Christoffels A, Clutterbuck D, Crowe M, Dalla E, Dalrymple B, de Bono B, Della Gatta G, di Bernardo D, Down T, Engstrom P, Fagiolini M, Faulkner G, Fletcher C, Fukushima T, Furuno M, Futaki S, Gariboldi M, Georgii-Hemming P, Gingeras T, Gojobori T, Green R, Gustincich S, Harbers M, Hayashi Y, Hensch T, Hirokawa N, Hill D, Huminiecki L, Iacono M, Ikeo K, Iwama A, Ishikawa T, Jakt M, Kanapin A, Katoh M, Kawasawa Y, Kelso J, Kitamura H, Kitano H, Kollias G, Krishnan S, Kruger A, Kummerfeld S, Kurochkin I, Lareau L, Lazarevic D, Lipovich L, Liu J, Liuni S, McWilliam S, Madan Babu M, Madera M, Marchionni L, Matsuda H, Matsuzawa S, Miki H, Mignone F, Miyake S, Morris K, Mottagui-Tabar S, Mulder N, Nakano N, Nakauchi H, Ng P, Nilsson R, Nishiguchi S, Nishikawa S (2005) The transcriptional landscape of the mammalian genome. Science 309:1559–1563
11. Amaral P, Mattick J (2008) Noncoding RNA in development. Mamm Genome 19:454–492
12. Lucchesi J, Kelly W, Panning B (2005) Chromatin remodeling in dosage compensation. Annu Rev Genet 39:615–651
13. Lee J (2010) The X as model for RNA's niche in epigenomic regulation. Cold Spring Harb Perspect Biol 2:a003749
14. Wutz A (2003) RNAs templating chromatin structure for dosage compensation in animals. Bioessays 25:434–442
15. Lyon MF (1961) Gene action in the X-chromosome of the mouse (Mus musculus L.). Nature 190:372–373
16. Brown C, Lafreniere R, Powers V, Sebastio G, Ballabio A, Pettigrew A, Ledbetter D, Levy E, Craig I, Willard H (1991) Localization of the X inactivation centre on the human X chromosome in Xq13. Nature 349:82–84
17. Rastan S, Robertson EJ (1985) X-chromosome deletions in embryo-derived (EK) cell lines associated with lack of X-chromosome inactivation. J Embryol Exp Morphol 90:379–388
18. Cattanach B, Isaacson J (1967) Controlling elements in the mouse X chromosome. Genetics 57:331–346
19. Kay GF, Barton SC, Surani MA, Rastan S (1994) Imprinting and X chromosome counting mechanisms determine Xist expression in early mouse development. Cell 77:639–650
20. Lyon MF (1999) Imprinting and X-chromosome inactivation. Results Probl Cell Differ 25:73–90
21. Avner P, Heard E (2001) X-chromosome inactivation: counting, choice and initiation. Nat Rev Genet 2:59–67
22. Boumil RM, Lee JT (2001) Forty years of decoding the silence in X-chromosome inactivation. Hum Mol Genet 10:2225–2232
23. Namekawa SH, Payer B, Huynh KD, Jaenisch R, Lee JT (2010) Two-step imprinted X inactivation: repeat versus genic silencing in the mouse. Mol Cell Biol 30:3187–3205
24. Xu N, Donohoe ME, Silva SS, Lee JT (2007) Evidence that homologous X-chromosome pairing requires transcription and Ctcf protein. Nat Genet 39:1390–1396
25. Xu N, Tsai CL, Lee JT (2006) Transient homologous chromosome pairing marks the onset of X inactivation. Science 311:1149–1152

26. Bacher C, Guggiari M, Brors B, Augui S, Clerc P, Avner P, Eils R, Heard E (2006) Transient colocalization of X-inactivation centres accompanies the initiation of X inactivation. Nat Cell Biol 8:293–299
27. Edwards C, Ferguson-Smith A (2007) Mechanisms regulating imprinted genes in clusters. Curr Opin Cell Biol 19:281–289
28. Sleutels F, Barlow DP (2002) The origins of genomic imprinting in mammals. Adv Genet 46:119–163
29. Bartolomei MS, Ferguson-Smith AC (2011) Mammalian genomic imprinting. Cold Spring Harb Perspect Biol. 3(7). pii: a002592. doi: 10.1101/cshperspect.a002592
30. Lee JT (2003) Molecular links between X-inactivation and autosomal imprinting: X-inactivation as a driving force for the evolution of imprinting? Curr Biol 13(6):R242–54
31. Cai X, Cullen B (2007) The imprinted H19 noncoding RNA is a primary microRNA precursor. RNA 13:313–316
32. Kevin V, M. (2009) RNA-directed control of transcription in human cells: specifically turning genes ON or OFF. Gene Ther Reg 04
33. Weinberg M, Morris K (2013) Long non-coding RNA targeting and transcriptional de-repression. Nucleic Acid Ther 23:9–14
34. Leung AK, Sharp PA (2006) Function and localization of microRNAs in mammalian cells. Cold Spring Harb Symp Quant Biol 71:29–38
35. Turner AM, Morris KV (2010) Controlling transcription with noncoding RNAs in mammalian cells. Biotechniques 48:ix–xvi
36. Ackley A, Lenox A, Stapleton K, Knowling S, Lu T, Sabir KS, Vogt PK, Morris KV (2013) An algorithm for generating small RNAs capable of epigenetically modulating transcriptional gene silencing and activation in human cells. Mol Ther Nucleic Acids 2:e104
37. Hawkins P, Santoso S, Adams C, Anest V, Morris K (2009) Promoter targeted small RNAs induce long-term transcriptional gene silencing in human cells. Nucleic Acids Res 37:2984–2995
38. Kim DH, Villeneuve LM, Morris KV, Rossi JJ (2006) Argonaute-1 directs siRNA-mediated transcriptional gene silencing in human cells. Nat Struct Mol Biol 13:793–797
39. Suzuki K, Juelich T, Lim H, Ishida T, Watanebe T, Cooper DA, Rao S, Kelleher AD (2008) Closed chromatin architecture is induced by an RNA duplex targeting the HIV-1 promoter region. J Biol Chem 283:23353–23363
40. Weinberg MS, Villeneuve LM, Ehsani A, Amarzguioui M, Aagaard L, Chen ZX, Riggs AD, Rossi JJ, Morris KV (2006) The antisense strand of small interfering RNAs directs histone methylation and transcriptional gene silencing in human cells. RNA 12:256–262
41. Han J, Kim D, Morris K (2007) Promoter-associated RNA is required for RNA-directed transcriptional gene silencing in human cells. Proc Natl Acad Sci U S A 104:12422–12427
42. Knowling S, Morris KV (2011) Epigenetic regulation of gene expression in human cells by noncoding RNAs. Prog Mol Biol Transl Sci 102:1–10
43. Hawkins P, Morris K (2008) RNA and transcriptional modulation of gene expression. Cell Cycle 7:602–607
44. Johnsson P, Ackley A, Vidarsdottir L, Lui WO, Corcoran M, Grander D, Morris KV (2013) A pseudogene long-noncoding-RNA network regulates PTEN transcription and translation in human cells. Nat Struct Mol Biol 20(4):440–446
45. Braunschweig M, Jagannathan V, Gutzwiller A, Bee G (2012) Investigations on transgenerational epigenetic response down the male line in F2 pigs. PLoS One 7:e30583
46. Janowski B, Younger S, Hardy D, Ram R, Huffman K, Corey D (2007) Activating gene expression in mammalian cells with promoter-targeted duplex RNAs. Nat Chem Biol 3: 166–173
47. Li LC, Okino ST, Zhao H, Pookot D, Place RF, Urakami S, Enokida H, Dahiya R (2006) Small dsRNAs induce transcriptional activation in human cells. Proc Natl Acad Sci U S A 103:17337–17342
48. Morris KV, Santoso S, Turner AM, Pastori C, Hawkins PG (2008) Bidirectional transcription directs both transcriptional gene activation and suppression in human cells. PLoS Genet 4: e1000258
49. Schwartz JC, Younger ST, Nguyen NB, Hardy DB, Monia BP, Corey DR, Janowski BA (2008) Antisense transcripts are targets for activating small RNAs. Nat Struct Mol Biol 15:842–848
50. Yu W, Gius D, Onyango P, Muldoon-Jacobs K, Karp J, Feinberg AP, Cui H (2008) Epigenetic silencing of tumour suppressor gene p15 by its antisense RNA. Nature 451:202–206
51. Hawkins PG, Morris KV (2010) Transcriptional regulation of Oct4 by a long non-coding RNA antisense to Oct4-pseudogene 5. Transcription 1:165–175
52. Modarresi F, Faghihi M, Lopez-Toledano M, Fatemi R, Magistri M, Brothers S, van der Brug M, Wahlestedt C (2012) Inhibition of natural antisense transcripts in vivo results in gene-specific transcriptional upregulation. Nat Biotechnol 30:453–459

Chapter 2

Isolation of Chromatoid Bodies from Mouse Testis as a Rich Source of Short RNAs

Oliver Meikar and Noora Kotaja

Abstract

RNA–protein (RNP) complexes and granules are powerful composites of merged functions and unique properties. The importance of RNPs in carrying out complex tasks in RNA processing and regulation is being increasingly revealed. One of the biggest RNP granules is the chromatoid body (CB) that is believed to orchestrate the RNA posttranscriptional regulation in haploid male germ cells. Here, we describe the CB isolation procedure, from mouse testis. After cross-linking and lysing the cells, the CBs are enriched by slow-speed centrifugation and immunoprecipitated using anti-MVH/DDX4 antibody. The method yields pure fractions of CBs, and it is robust, reproducible and does not require special equipment or abundant starting material. The CB is packed with large amounts of RNA, especially small RNAs. Isolation of the CBs provides a tool to enrich these RNA species.

Key words Chromatoid body, Germ granule, Germ cell, Round spermatids, Spermatogenesis, Ribonucleoprotein complex, Immunoprecipitation, PIWI-interacting RNA

1 Introduction

RNA and protein can combine to form ribonucleoprotein (RNP) particles that serve various functions. Known examples of RNP complexes include ribosomes, telomerase, heterogeneous nuclear RNPs (hnRNP), small nuclear RNPs (snRNP), and RNA-induced silencing complex (RISC). RNPs can form larger granules that represent a highly versatile group of cellular structures with unique features and vital roles. These RNP granules, such as stress granules, processing bodies, and exosomes, are often involved in RNA posttranscriptional regulation, for example in the translational control of mRNAs [1–4]. One of the largest RNP granules is the chromatoid body (CB) in haploid male germ cells [5, 6]. The CB is suggested be a central structure for RNA posttranscriptional regulation during haploid cell differentiation into mature sperm. This is a complex process characterized by extensive transcription with delayed or lacking translation and high requirements for the

Andreas Werner (ed.), *Animal Endo-siRNAs: Methods and Protocols*, Methods in Molecular Biology, vol. 1173, DOI 10.1007/978-1-4939-0931-5_2,

accurate quality control and surveillance of transcripts. The CB consists of large amounts of RNA-binding proteins and different classes of RNAs [5]. The predominant RNA species in the CB are small RNAs, mostly PIWI-interacting RNAs (piRNAs) [5, 7, 8]. Because of its size of around 1 μm, the CB is easily visible with phase-contrast microscopy and has therefore been known for over a century. Different approaches such as gradient centrifugation have been applied for the CB isolation but only with limited success [9].

Immunoprecipitation (IP) is a powerful technique to enrich the antigen from a heterogeneous solution using a respective antibody against that antigen. The antibody itself is attached to a carrier, for example a paramagnetic bead that allows for washing and separating the bound antigen from the rest of the sample mix. The most important parameter of the successful IP is the specificity and sensitivity of the used antibody.

The isolation procedure described here takes advantage of the compact nature and the big size of the CB and an excellent antibody against its major protein component MVH/DDX4 (Mouse Vasa Homologue) [10] (Fig. 1). The starting material is a mixed population of male germ cells released from the mouse testes by collagenase digestion. The cells are cross-linked with formaldehyde to preserve the molecular interactions inside the CB. Cell lysis with detergents and by sonication is followed by a centrifugation step

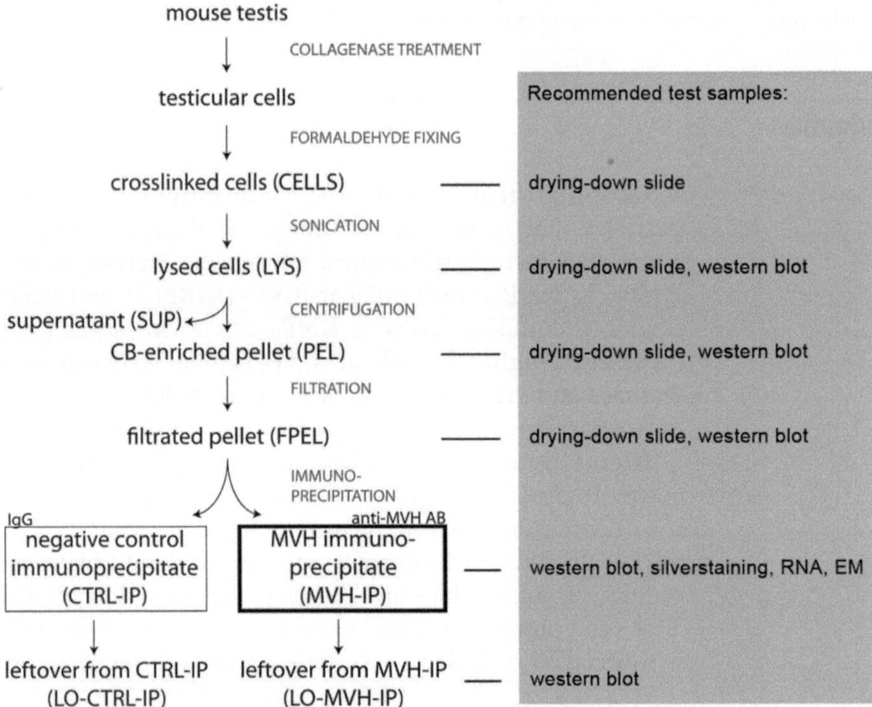

Fig. 1 Schematic diagram of the chromatoid body isolation from mouse testis. To monitor the success of the isolation, we suggest the analysis of test samples from different steps of the protocol

that concentrates the CBs and removes the non-CB associated MVH. The CB-enriched lysate is then subjected to IP with a commercially available anti-MVH antibody.

To monitor the success of the CB isolation, take test samples from different steps during the procedure as indicated in the protocol (Fig. 1). Western blot analysis best reflects the efficiency of the CB isolation and requires only small amounts of the sample (Fig. 2b). Immunostaining of drying-down slide samples allows monitoring the integrity and yields of CBs during the enrichment steps (the protocol for drying-down slides is provided in Subheading 3.6) (Fig. 2a). The analysis of RNA samples requires ten times more material and is not informative in monitoring cell lysis and CB enrichment steps. However, the RNA profile of the isolated CBs is characteristic (Fig. 2d) and should be analyzed before any further applications with CB-RNA. The best method to study the integrity and purity of the isolated CBs is to analyze the immunoprecipitated CB-beads by transmission electron microscopy (Fig. 2c).

2 Materials

Prepare all solutions using Milli-Q-filtered and deionized (MQ) water. You will need 1.5 ml microcentrifuge tubes and 50 ml falcon tubes and pipettes to pipet 1 μl–5 ml.

2.1 Working Solutions

Prepare the following buffers before you start. All reagents can be stored for at least a month at room temperature unless otherwise indicated.

1. 1× phosphate buffered saline (PBS): 137 mM NaCl, 2.7 mM KCl, 10 mM Na_2HPO_4, 1.8 mM KH_2PO_4, pH 7.4.
2. 10× glycine solution: 2.5 M glycine. Add 9.37 g of glycine to 40 ml of MQ water. Adjust the pH to 7.0 with sodium hydroxide and make it up to 50 ml.
3. RIPA buffer: 50 mM Tris–HCl (pH 7.5), 1 % NP-40, 0.5 % w/v sodium deoxycholate, 0.05 % w/v sodium dodecyl sulfate, 1 mM EDTA, 150 mM NaCl. For 200 ml of stock solution, mix together 10 ml of 1 M Tris–HCl, 2 ml NP-40, 1 g sodium deoxycholate, 1 ml 10 % SDS, 1 ml 0.2 M EDTA, and 6 ml 5 M NaCl. The pH will be around 7.3–7.4. Add protease inhibitors just before use.
4. Citrate–phosphate buffer: 24.5 mM citric acid, 51.7 mM dibasic sodium phosphate. Measure 4.7 g of citric acid, 9.2 g of $Na_2HPO_4 \cdot 2H_2O$ and fill up to 1 l with MQ water. Adjust pH to 5.0.
5. Citrate–phosphate buffer + 0.1 % Tween: Add 0.5 ml Tween-20 to 0.5 l of citrate–phosphate buffer described above.

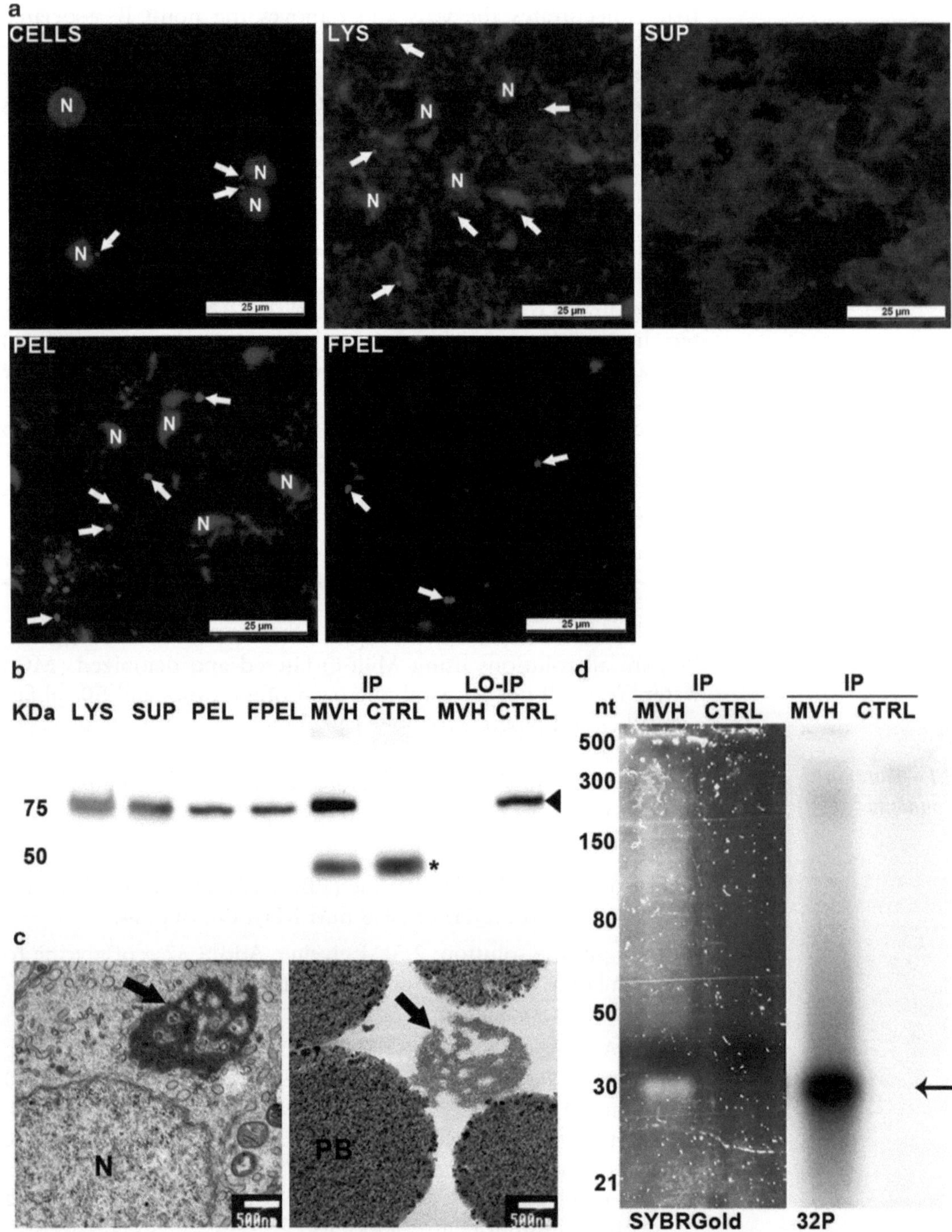

Fig. 2 Validation of the chromatoid body isolation. (**a**) Immunofluorescent staining of drying-down slide preparations representing different steps of the chromatoid body isolation procedure. *CELLS* cross-linked cells, *LYS* cell lysate, *SUP* supernatant fraction of the centrifuged cell lysate, *PEL* pellet of the centrifuged cell lysate, *FPEL* filtered pellet fraction. *N* nuclei of testicular cells. Chromatoid bodies are indicated with *arrows*. Scale bars: 25 μm. (**b**) Western blot analysis tracing MVH (indicated by *arrowhead*) in test samples at different steps of the chromatoid body isolation procedure. *MVH-IP* immunoprecipitate with anti-MVH antibody.

6. 50× Complete Mini Protease Inhibitor without EDTA (Roche): Dissolve one tablet in 1 ml MQ water. Can be stored at −20 °C for months.
7. 1 M DTT: Dissolve 0.154 g of DTT in 1 ml MQ water. Can be stored at −20 °C for months.
8. 0.2 M PMSF: Dissolve 0.035 g of PMSF in 1 ml isopropanol. Can be stored at −20 °C for months. May need mild heating for dissolving the crystals.
9. Drying-down fixing solution: 0.15 % TritonX-100, 1 % PFA in PBS. Mix 305 μl PBS, 4.8 μl 10 % TritonX-100 and 10 μl 32 % PFA. Can be stored at −20 °C for at least a month.
10. S-collidine buffer (0.16 M): Dissolve 2.136 g of pure s-collidine in 80 ml of water, adjust pH to 7.4 with 1 N HCl and full up to 100 ml.

2.2 Preparing the Cells from Mouse Testis Tissue

1. Forceps and scissors to cut, decapsulate, and mince mouse testes.
2. Collagenase solution: 0.05 % (w/v) collagenase and 0.1 % (w/v) glucose in 1× PBS. Add 20 mg collagenase (Worthington) and 40 mg glucose to 40 ml of 1× PBS. Prepare in 50 ml Falcon tube immediately before use.
3. 100 μm cell strainer (BD Falcon) for filtering the collagenase treated testis.
4. Vertical rotator for 50 ml centrifuge tubes and 1.5 ml microcentrifuge tubes (for example Multi RS-60, Biosan). Use medium speed (e.g., 10 turns/min).
5. 32 % PFA (Electron Microscopy Sciences, USA). Open the fresh vial on the day of use and keep on ice.
6. Centrifuge with a rotor fitting 50 ml centrifuge tubes, allowing centrifugation at 300 × *g* at 4 °C for the washing of the cells.
7. 10× glycine solution for quenching the cross-linking reaction (*see* **item 2** of Subheading 2.1).
8. Humidified chamber and glass slides with circled grease pen sections, for the preparation of drying-down slides.

Fig. 2 (continued) *CTRL-IP* negative control immunoprecipitate with anti-acrosin antibody. *LO-IP* leftover fraction after immunoprecipitation. IgG heavy chain is indicated by *asterisk*. (**c**) Transmission electron microscopy images of the chromatoid body (indicated by *arrow*) in a spermatid next to a nucleus (*N*) and after the immunoprecipitation, attached on paramagnetic beads (*PB*). Scale bars: 500 nm. (**d**) RNA profile of the chromatoid body. 5′ radiolabeled RNA is separated in denaturing 15 % PAA gel and visualized by SYBRGold staining and phosphorimaging. *Arrow* indicates the characteristic piRNA band

2.3 Cell Lysis and the Enrichment of the CB Fraction

1. Just before lysing the cells, add protease inhibitors to RIPA buffer: per 5 ml of RIPA buffer (*see* **item 3** of Subheading 2.1), add 100 μl of 50× complete mini EDTA-free solution (*see* **item 6** of Subheading 2.1), 5 μl 0.2 M PMSF (*see* **item 8** of Subheading 2.1), 5 μl 1 M DTT (*see* **item 7** of Subheading 2.1) to get to the final concentration of 1× complete mini mix, 0.2 mM PMSF, and 1 mM DTT. Put on ice.
2. Ice box with ice to keep the temperature low during cell lysis.
3. Sonicator (UCD-200, Diagenode) for 1.5 ml microcentrifuge tubes.
4. Tabletop centrifuge allowing centrifugation of 1.5 ml microcentrifuge tubes at 300 × g at 4 °C for pelleting the CB fraction.
5. Millex-SV 5.00 μm filter units (SLSV 025LS, Millipore).
6. Two 2 ml syringes for the filter units.

2.4 Preparing Antibody–Dynabeads Complexes

1. Magnetic rack for 1.5 ml eppendorf tubes.
2. Dynabeads protein G (Invitrogen) (*see* **Note 1**).
3. Anti-MVH (DDX4) antibody. We have successfully used in-house generated rabbit anti-MVH and commercial rabbit polyclonal anti-MVH antibody (Abcam, ab13840).
4. Negative Control for Rabbit IgG antibody (Neomarkers, NC-100-P) or a non-CB specific antibody, e.g., anti-acrosin (Santa Cruz, sc-67151).
5. Citrate–phosphate buffer for protein G—antibody binding (*see* **item 4** of Subheading 2.1) (*see* **Note 2**).
6. Citrate–phosphate buffer + 0.1 % Tween for washing the antibody–beads (*see* **item 5** of Subheading 2.1).

2.5 Reversing the Cross-links and the Analysis of the Results

1. Heat block at 70 °C.
2. Trizol reagent or other phenol-based RNA extraction reagent for isolating the CB-RNA (*see* **Note 3**).
3. Standard RNA electrophoresis equipment for 15–20 % denaturing urea polyacrylamide gel analysis.
4. SYBRGold for visualizing the RNA in the gel.
5. Standard western blot equipment.
6. 2× Laemmli buffer: 4 % SDS, 20 % glycerol, 10 % 2-mercaptoethanol, 0.004 % bromophenol blue, 0.125 M Tris–HCl, pH 6.8.
7. Anti-MVH antibody for western blot and drying-down slide analysis (*see* **Note 4**).
8. Silver staining kit for visualizing the CB proteome.
9. 5 % glutaraldehyde in 0.16 M s-collidine buffer (**item 10** of Subheading 2.1), pH 7.4 for preparing the electron microscopy sample.

3 Methods

Carry out all procedures at room temperature unless otherwise specified. In order to monitor the steps of the CB enrichment and isolation, take test samples as indicated in the protocol (Fig. 1). Use following volumes for the tests samples:

Drying-down slides: 5 μl (*see* protocol in Subheading 3.6).

Western blot samples: 5 μl. Mix with 5 μl of Laemmli buffer, reverse the cross-links (*see* Subheading 3.5).

RNA PAGE analysis: 50 μl. Reverse the cross-links (*see* Subheading 3.5) and isolate the RNA (*see* **Note 3**).

Timeline:

Day 1: Liberate the cells from testis tissue, cross-link, and lyse. Enrich the CB fraction from the cell lysate. At the same time prepare the Dynabeads–antibody complexes (4 h).

Preclear the enriched CB fraction with unbound Dynabeads (1 h). Start the immunoprecipitation and incubate overnight (*see* **Note 5**).

Day 2: After the immunoprecipitation, wash the beads and reverse the cross-links (1.5 h).

Continue with other downstream analyses.

3.1 Preparation of the Testicular Cells

1. Sacrifice two sexually mature mice and dissect out the testes avoiding the epididymides.
2. Decapsulate the testes and mince the seminiferous tubules with scissors ten times.
3. Immediately transfer the tubules into falcon tubes with 40 ml of collagenase solution (*see* **Note 6**).
4. Digest the seminiferous tubules for 1 h on a vertical rotator. Shake occasionally to help the seminiferous tubules to entangle (*see* **Note 7**).
5. Meanwhile, prepare the Dynabeads–antibody complexes (*see* Subheading 3.3).
6. Pass the cell suspension through a 100 μm cell strainer to remove remaining tubule fragments from liberated cells.
7. Pellet the cells at 300 × *g* for 5 min at 4 °C.
8. Remove the supernatant and resuspend the pellet in 40 ml of cold PBS (*see* **Note 8**).
9. Pellet the cells again at 300 × *g* for 5 min at 4 °C, remove the supernatant. The pellet contains the liberated and washed testicular cells. Resuspend the cells in 20 ml of PBS.

10. For cross-linking, add 62.5 μl of fresh 32 % PFA to the resuspended cells to obtain a 0.1 % PFA solution. Mix immediately by inverting the tube and incubate on the vertical rotator for 20 min (*see* **Note 9**).
11. Stop the cross-linking reaction by adding 2 ml of 10× glycine solution and incubate for 5 min on the vertical rotator.
12. You may now take the first drying-down slide sample for monitoring the CB isolation process (Sample name: CELLS).
13. Pellet the fixed cells by centrifuging at 300×*g* for 5 min at 4 °C and resuspend the pellet in 40 ml PBS (*see* **Note 8**).
14. Pellet the cells again at 300×*g* for 5 min at 4 °C and remove all the supernatant (*see* **Note 10**).

3.2 Cell Homogenization and the Enrichment of the CB Fraction

1. Resuspend the fixed cells in RIPA buffer in a total volume of 1.5 ml, on ice.
2. Divide the suspension into 6×1.5 ml microcentrifuge tubes (250 μl each).
3. Sonicate the tubes in an ice-cold water-bath at medium setting for 6×30 s with 30 s intervals.
4. Transfer the tubes on ice.
5. You may now take the test samples of the lysed cells. Sample name: LYS (*see* **Note 11**).
6. Centrifuge the tubes with lysed cells at 300×*g* for 10 min at 4 °C and immediately remove the supernatant fraction with a pipette, before the CB-enriched pellet detaches from the bottom of the tubes (*see* **Note 12**).
7. You may now take the test samples of the supernatant of the lysed cells. Sample name: SUP.
8. Resuspend the pellets by flicking the tubes and then adding 250 μl of RIPA buffer to each of the six tubes.
9. To facilitate the resuspension, sonicate the tubes again at medium setting for 2×30 s, with 30 s intervals (*see* **Note 13**).
10. Pool the content of the six tubes together and put on ice.
11. You may now take the test samples of the pellet fraction of the lysed cells. Sample name: PEL.
12. Using the syringe, filter the pooled suspension a through 5.0 μm filter to remove the sperm heads and other larger structures that co-sediment during the centrifugation step (*see* **Note 14**).
13. Flush any retained sample from the filter with an additional ~200 μl of RIPA buffer using the second syringe and empty by air-purge.

14. You may now take a test sample of the filtrated pellet fraction of the lysed cells. Sample name: FPEL (*see* **Note 15**). You have now about 1–1.4 ml of enriched CB solution that is ready for immunoprecipitation.

3.3 Preparing Dynabeads–AB Complexes

Start preparing the beads during the 1 h collagenase digestion of the seminiferous tubules (*see* **Note 16**).

1. Resuspend the Dynabeads Protein G thoroughly (*see* **Note 17**).
2. Transfer 45 μl of the beads to a 1.5 ml microcentrifuge tube.
3. Place the tube on the magnet to separate the beads from solution and discard the supernatant (*see* **Note 18**).
4. Remove the tube from the magnet and resuspend the beads in 1.5 ml of citrate–phosphate buffer.
5. Divide the diluted beads solution equally between three tubes, 500 μl each.
   (a) 500 μl beads for preclearing the CB-enriched sample before immunoprecipitation.
   (b) 500 μl beads for conjugating with anti-MVH antibody.
   (c) 500 μl beads for conjugating with the control antibody.
6. Add approximately 8 μg of anti-MVH to sample B and 8 μg of control antibody to sample C.
7. Incubate the tubes on a vertical rotator for a minimum of 30 min to couple the antibodies to the beads via protein G.
8. Collect the antibody–bead complexes with the magnet and discard the supernatants.
9. Remove the tubes from the magnet and wash the beads shortly with 1 ml of citrate–phosphate buffer + 0.1 % Tween.
10. Repeat the washing (**steps 8** and **9**) once.
11. Keep the washed bead–antibody complexes in RIPA buffer until you have precleared your filtrated cell pellet fraction and then proceed with immunoprecipitation (*see* Subheading 3.4).

3.4 Immuno-precipitating the CBs

1. Before the immunoprecipitation, preclear the filtrated cell pellet fraction using uncoupled Dynabeads. Place the tube with preclearing beads (*see* **step 5**(a) of Subheading 3.3) on a magnet and remove the buffer.
2. Add the filtered pellet fraction (*see* **step 14** of Subheading 3.2) to the preclearing beads.
3. Incubate on a vertical rotator at least for 30 min.
4. Place the tube on a magnet to collect the preclearing beads.
5. Measure the amount of the precleared sample while pipetting it to a new tube, because you need to divide it equally between the MVH- and control-antibody beads.

6. Place the tubes with washed bead–antibody complexes (*see* **step 11** of Subheading 3.3) on a magnet and remove the buffers.
7. Immediately divide the precleared sample equally between both tubes with antibody–bead complexes and start the immunoprecipitation (MVH-IP and CTRL-IP).
8. Incubate with rotation overnight at 4 °C to allow the CBs to bind to the antibody–bead complexes (*see* **Note 5**).
9. After immunoprecipitation of the CBs, place the IP tubes on a magnet to separate the antigen–antibody–Dynabeads complexes (IP-beads) from the solution.
10. You may now take test samples of the supernatants (Left-Over-samples) that are depleted from the IP-beads, to monitor the success of the immunoprecipitation. Sample names: LO-MVH-IP and LO-CTRL-IP (*see* **Note 19**).
11. Wash the IP-beads by carefully suspending them in 750 μl of RIPA buffer and incubating on a vertical rotator for 10 min at RT.
12. Place the tubes on a magnet to collect the washed IP-beads and remove the supernatant.
13. Repeat the washing **steps 10** and **11** two times.
14. During the last wash step you can take the most important test samples of the isolated CBs and the control IP. Sample names: MVH-IP and CTRL-IP. Keep the volume same in **step 10**.
15. You have purified the CBs from the testes of two mice and obtained CB-associated RNA and CB-associated proteins roughly in hundreds of nanograms (For scaling up the protocol, *see* **Note 20**).

3.5 Preparation of the CB-IP Samples for Analysis

For most of the downstream applications, you need to reverse the cross-links. For maximum yield reverse the cross-links while the CBs are still bound to the AB-bead complexes.

1. Reverse the PFA cross-links by heating the samples at 70 °C, 45 min in the appropriate buffer (see below):
    - For RNA isolation, resuspend the IP-beads in 50–100 μl of RIPA buffer, reverse the cross-links and proceed with RNA isolation (*see* **Note 3**). Dissolve the isolated RNA in 10–20 μl of nuclease-free water. Common photospectrometers at 260 nm are not sensitive enough to measure the isolated CB RNA, but indicates the purity of the sample. Use Qubit (Invitrogen) for measuring the CB RNA concentration or a Bioanalyzer (Agilent) for further analysis. We use routinely 15–20 % denaturing urea polyacrylamide gel electrophoresis with SYBRGold

post-staining (1:10,000) to visualize CB RNA (Fig. 2d). One tenth of the purified CB RNA sample corresponding to tens of nanograms is enough to reveal the characteristic CB-RNA profile by PAGE. For detecting 1–2 orders lower RNA concentrations, 5′ radiolabel the CB RNA before PAGE analysis and detect by PhosphorImager system (Fig. 2d).

- For WB and silver staining analysis, dilute the CB-beads in 20 μl of RIPA buffer, add the same volume of 2× Laemmli buffer and reverse the cross-links. Proceed with the standard protocol for WB using 1/200 of the sample, and 1:1,000 of anti-MVH antibody (*see* **Note 4**) and the ECL detection system. For good quality silver staining results, apply the whole CB-IP sample to the commercial gradient gel and continue with the standard silver staining protocol.
- For MS analysis, run the whole CB-IP sample shortly in the commercial gradient gel (e.g., mini-protean TGX precast gel, Bio-Rad). Stain with Colloidal Blue staining kit to visualize the protein bands and carefully cut out the appropriate gel pieces from the MVH-IP and control-IP sample lanes, avoiding any cross-contamination. Proceed with in-gel digestion and the standard protocol for preparing the samples for MS analysis. Alternatively, if size separation is not required, the CB proteins can be purified and concentrated by other means (ultracentrifugal filters, precipitation) for MS analysis.
- For transmission electron microscopy analysis (Fig. 2c), do not reverse the cross-links. Pellet a small sample of IP-beads to the bottom of the conical electron microscopy sample tube and replace the liquid carefully with 5 % glutaraldehyde. Fix for at least 3 h and continue with the standard protocol for preparing the EM samples.

3.6 Preparation of Drying-Down Slides from the Samples

The drying-down technique was initially developed to visualize the synaptonemal complexes of meiotic cells [11]. The protocol has also been applied for the analysis of CBs in round spermatids (*see* **Note 21**) [12, 13].

1. Apply 10 μl of drying-down slide fixing solution (**item 9** of Subheading 2.1) inside the water repelling grease-pen circle on the slide.
2. Immediately add 5 μl of the sample taken at different steps of the CB isolation protocol. Mix the solutions gently by tilting the slide and incubate the slides in a highly humidified chamber overnight.
3. The next day, remove the slides from the chamber and let them air-dry until completely dry (*see* **Note 22**).

4. Post-fix the slides in 4 % paraformaldehyde in PBS for 5 min.
5. Remove the fixing solution and wash the slides three times with PBS.
6. Treat the slides with 0.2 % Triton-X-100 in PBS for 2 min.
10. Wash three times with PBS and then block with 10 % BSA in PBS for 30 min.
7. Incubate the slides with rabbit polyclonal anti-MVH antibody (1:300) for 1 h at room temperature or overnight at 4 °C.
8. Wash the slides three times 5 min with PBS + 0.1 % Tween-20.
9. Incubate with the secondary antibody (e.g., Alexa Fluor 594 anti-rabbit IgG, 1:500) for 1 h at room temperature.
10. Wash the slides three times for 5 min with PBS + 0.1 % Tween-20 followed by a rinse with PBS.
11. Mount in mounting medium with DAPI (4′,6-diamidino-2-phenylindole).

4 Notes

1. Instead of Protein G you can use other surface functionalities to covalently couple the primary antibody to the Dynabeads. However, in our hands other strategies to couple the antibody to the Dynabeads result in lower yields of immunoprecipitated CBs.
2. Alternatively, PBS can be used for antibody binding and washing steps.
3. Avoid purifying the CB RNA with spin columns as it may alter the integrity of your whole CB RNA profile. Use glycogen or linear acrylamide as a carrier when precipitating the RNA, because the quantities are low.
4. In addition to anti-MVH, you can use other CB-specific antibodies for monitoring the success of the immunoprecipitation, for example anti-MIWI or anti-DDX25 antibodies.
5. Most of the CBs bind to Dynabeads–antibody complexes within an hour and practically all the CBs are bound when incubated 8 h at room temperature or overnight at 4 °C.
6. Have the tube with collagenase solution with you when collecting and mincing the testes in order to continue with the collagenase digestion without delay.
7. Observe the seminiferous tubules during collagenase digestion. Well-digested tubules are filamentous and about half a centimeter long. If you see many big clumps of tubules, shake the falcon tube gently to enhance the digestion.

8. It is easiest to resuspend the pellets in a tiny volume of buffer, for example in the backflow from the emptying of the tube by flicking the tube.
9. Cross-linking is a crucial step of the method and cannot be bypassed. Without cross-linking, the CBs will dissociate during cell lysis. The concentration of PFA can be altered at least in the range of 0.05–0.2 %.
10. Try to remove as much liquid from the fixed cell pellet as possible because the extra liquid will dilute the RIPA buffer in the following lysis step. One way to get rid of the excess liquid after pouring away the supernatant is keeping the tube upside down for a minute and drying the leftover droplet with a tissue. Flick the tube and put on ice. With some backflow buffer, there should be around 100–300 μl of the cell pellet.
11. While the CB-associated RNA is protected from RNases, the non-CB associated RNA of the lysed cells gets degraded in minutes. If the LYS RNA sample is required, the best way to retain most of its integrity is to be very rigorous in keeping the cell lysate constantly ice-cold. Use ice-cold RIPA buffer for lysis and immediately after sonication, treat the sample with proteinase K by pipetting 50 μl of the sonicated cell lysate into 0.5 μl of Proteinase K (20 mg/ml), 2.5 μl 10 % SDS and incubating at 50 °C for 40 min.
12. This centrifugation is crucial for CB isolation. It takes advantage of the big size of CBs, leaving smaller particles, soluble matter and DNA in the supernatant phase. A range of at least 200–400 × *g* is suitable for the isolation of CBs.
13. When required, **steps 6–8** can be repeated to produce a more pure fraction of the cell pellet.
14. This filtration step is not required when immature or genetically modified mice lacking sperm are used or when sperm heads are not interfering with downstream analysis (e.g., CB functional assays).
15. Because you wash the filter with additional buffer, the FPEL fraction will be diluted. Consider this when comparing the different samples documenting the CB enrichment.
16. To optimize the antibody/bead ratio and coupling time, you can take 2 μl control samples from the beads and the supernatant with unbound antibodies during and after the coupling reaction of anti-MVH antibody with Dynabeads protein G. In the conditions described in this protocol, all the antibodies get bound to the beads.
17. Dynabeads are heavy and settle quickly at the bottom of the tube. Make sure that the beads are evenly solubilized when dividing them or when taking samples.

18. Never let the beads dry, always keep them in some solution.
19. The leftover of the control-IP is actually a precleared, CB-enriched cell pellet fraction. Instead of discarding it after taking the test samples, you can reuse the rest of LO-CTRL-IP for isolating CBs by just adding a new batch of anti-MVH–bead complexes and continuing with the immunoprecipitation (from **step 8** of Subheading 3.4).
20. When you feel safe with the method and want to scale it up, do so proportionally. The method is optimized for two mice per 50 ml collagenase tube, 45 μl of beads, six tubes for sonication, one 5 μm filter. Only scaling up the starting material does not necessarily produce higher yields.
21. Immunofluorescent staining of drying-down slide preparations is a reliable method to trace the CBs during the enrichment steps of this CB isolation protocol. However, the drying-down slide analysis is cumbersome because the immunoprecipitated CBs elute poorly from MVH-Dynabeads. Moreover, the autofluorescent properties of Dynabeads interfere with the immunofluorescent analysis of the uneluted CB-bead complexes.
22. At this point, the slides can be stored at room temperature for at least several days.

Acknowledgements

This work was supported by the Academy of Finland, Emil Aaltonen Foundation, and Turku Doctoral Programme of Biomedical Sciences.

References

1. Stefl R, Skrisovska L, Allain FH (2005) RNA sequence- and shape-dependent recognition by proteins in the ribonucleoprotein particle. EMBO Rep 6:33–38
2. Anderson P, Kedersha N (2006) RNA granules. J Cell Biol 172:803–808
3. Anderson P, Kedersha N (2009) RNA granules: post-transcriptional and epigenetic modulators of gene expression. Nat Rev Mol Cell Biol 10:430–436
4. Balagopal V, Parker R (2009) Polysomes, P bodies and stress granules: states and fates of eukaryotic mRNAs. Curr Opin Cell Biol 21:403–408
5. Meikar O, Da Ros M, Korhonen H, Kotaja N (2011) Chromatoid body and small RNAs in male germ cells. Reproduction 142:195–209
6. Kotaja N, Sassone-Corsi P (2007) The chromatoid body: a germ-cell-specific RNA-processing centre. Nat Rev Mol Cell Biol 8:85–90
7. Meikar O, Da Ros M, Liljenback H, Toppari J, Kotaja N (2010) Accumulation of piRNAs in the chromatoid bodies purified by a novel isolation protocol. Exp Cell Res 316:1567–1575
8. Klattenhoff C, Theurkauf W (2008) Biogenesis and germline functions of piRNAs. Development 135:3–9
9. Figueroa J, Burzio LO (1998) Polysome-like structures in the chromatoid body of rat spermatids. Cell Tissue Res 291:575–579
10. Toyooka Y, Tsunekawa N, Takahashi Y, Matsui Y, Satoh M, Noce T (2000) Expression and intracellular localization of mouse Vasa-homologue

protein during germ cell development. Mech Dev 93:139–149

11. Peters AH, Plug AW, van Vugt MJ, de Boer P (1997) A drying-down technique for the spreading of mammalian meiocytes from the male and female germline. Chromosome Res 5:66–68

12. Kotaja N, Kimmins S, Brancorsini S, Hentsch D, Vonesch JL, Davidson I, Parvinen M, Sassone-Corsi P (2004) Preparation, isolation and characterization of stage-specific spermatogenic cells for cellular and molecular analysis. Nat Methods 1:249–254

13. Kotaja N, Bhattacharyya SN, Jaskiewicz L, Kimmins S, Parvinen M, Filipowicz W, Sassone-Corsi P (2006) The chromatoid body of male germ cells: similarity with processing bodies and presence of Dicer and microRNA pathway components. Proc Natl Acad Sci U S A 103: 2647–2652

Chapter 3

Generation of Endo-siRNAs in *Xenopus laevis* Oocytes

Sammer Alnumeir and Andreas Werner

Abstract

Endogenous siRNAs (endo-siRNAs) are well documented and characterized in *C. elegans* and *Drosophila*. Endo-siRNAs can also be found in vertebrates; however, their biology is much less clear. They are thought to be produced by Dicer and to contribute to transposon silencing. Because of their generally low abundance and their similarity with miRNAs and products of physiological RNA turn-over, endo-siRNAs are difficult to investigate. Here, we report a system, oocytes from *Xenopus laevis*, that allows for the generation and analysis of endo-siRNAs from double-stranded RNA precursors.

Key words Endo-siRNA, *Xenopus laevis* oocytes, Nuclear injections

1 Introduction

Endo-siRNAs derive from double-stranded RNA precursors formed by complementary transcripts or internal hairpins. Significant sources of endo-siRNAs are bidirectionally transcribed transposons, pseudogenes, and natural sense–antisense pairs [1]. How these precursors can be generated in significant amounts is still unclear. Vertebrates lack an enzyme that generates double-stranded RNA from single-stranded precursors (RNA-dependent RNA polymerase, RdRP) [2]; moreover, significant levels of double-stranded RNA in the cytoplasm of vertebrate cells trigger an immune response via the activation of protein kinase R (PKR) and interferon-γ [3]. As a consequence, endo-siRNAs were initially believed to be restricted to lower eukaryotes. The increased depth of parallel sequencing developed in recent years, however, revealed that significant levels of endo-siRNAs can be detected—at least in selected tissues [4–6].

The biological phenomena associated with the formation of double-stranded RNA, transposon mobilization, and the expression of natural antisense transcripts show development and tissue specific expression. The findings that germ cells as well as embryonic cells

Andreas Werner (ed.), *Animal Endo-siRNAs: Methods and Protocols*, Methods in Molecular Biology, vol. 1173,
DOI 10.1007/978-1-4939-0931-5_3, © Springer Science+Business Media New York 2014

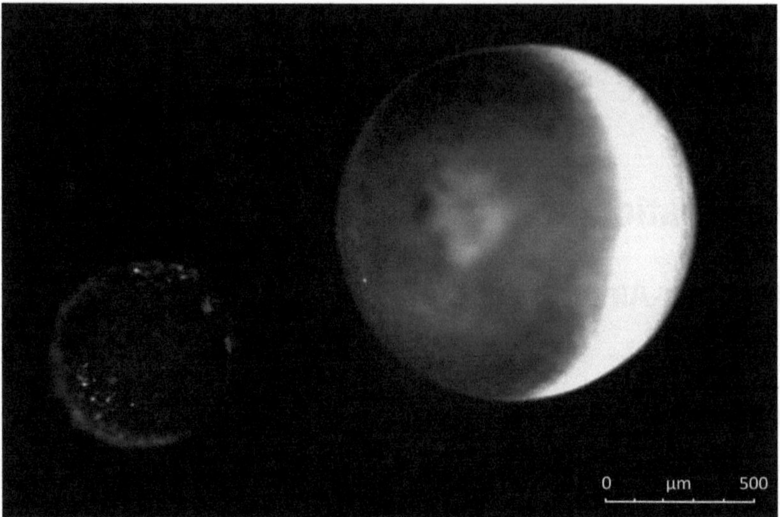

Fig. 1 *Xenopus* oocyte and an isolated nuclear envelope

show pervasive transcription and reduced DNA methylation provide circumstantial evidence that endo-siRNAs may be produced during these developmental stages [7]. Accordingly, endo-siRNAs have been identified in mouse testis and oocytes [5]. We have found that fully complementary sense–antisense RNAs injected into the nuclei of *Xenopus laevis* oocytes are processed into endo-siRNAs. Mismatches between the RNA strands are not tolerated, and the region of complementarity needs to be at least 30 base pairs [8]. If the complementary RNAs are injected into the cytoplasm of the oocytes, both transcripts remain stable for days, protected by a double-stranded RNA binding protein [9].

In this article we describe a robust strategy how to generate short RNAs reminiscent of endo-siRNAs using the *Xenopus laevis* oocyte expression system (Fig. 1). We also discuss the different possibilities to assess RNAi related processes using the system and mention its limitations.

2 Materials

Prepare all solutions using distilled water and autoclave the stock solutions. Store the stock solutions at 4 °C and the ORII and Barth's solution at 18 °C. Perform all the reactions in an RNase-free environment when producing the RNA for injection.

Animals: *Xenopus laevis* can be bread in house according to local guidelines. The Web site http://www.xenbase.org/common/ contains a wealth of useful tips and contacts. Various strategies to harvest oocytes exist and have been published in detail [10, 11].

In addition, there are commercial services that isolate and ship injection ready oocytes (EcoCyte Bioscience).

2.1 Equipment

1. Microinjector system suitable for volumes between 1 and 50 nL. Numerous systems exist with varying degrees of automatization, and the prices vary accordingly.
2. Glass capillary puller, a very basic model will do.
3. Cell culture dish to keep the oocytes immobilized and submersed in Barth's solution for injections. The easiest way is to scratch the plates with a blade to stop the oocytes from escaping the injection needle. Alternatively, a nylon mesh can be glued to the dish with nail varnish.
4. Optional: Micro plate centrifuge.
5. Optional: Mini multiwell culture plates with conical wells (Terasaki plates; Nunc or Greiner, for example).

2.2 Solutions for Xenopus Oocyte Experiments

1. Modified Barth's solution (Storage solution): The composition is 88 mM NaCl, 1 mM KCl, 0.41 mM $CaCl_2$, 0.82 mM $MgSO_4$, 0.33 mM $Ca(NO_3)_2$, 7.5 mM HEPES, 2.4 mM $NaHCO_3$. We use autoclaved stock solutions of all these components (5 M NaCl [17.6 mL], 1 M KCl, [1 mL], 100 mM $CaCl_2$ [4.1 mL], 100 mM $MgSO_4$ [8.2 mL], 1 M $Ca(NO_3)_2$ [0.33 mL], 1 M $NaHCO_3$ [2.4 mL], 1 M HEPES [7.5 mL]), add the aliquots to 800 mL of water, adjust to pH 7.5 with Tris base and autoclave. Add gentamycin to a final concentration of 0.02 mg/mL the cold solution and store at 18 °C.
2. ORII- Ca^{+2}-free solution: The composition is 82.5 mM NaCl, 2 mM KCl, 1 mM $MgCl_2$, 10 mM HEPES. Add the aliquots as above [16.5 mL NaCl stock, 2 mL KCl stock, 1 mL of 1 M $MgCl_2$ stock, 10 mL of HEPES stock], adjust to pH 7.5 with Tris base, autoclave, and store at 18 °C.
3. Collagenase A, 2 mg/mL in ORII.

2.3 RNAs for Xenopus Oocyte Injections

In vitro synthesized RNA is usually injected into the cytoplasm and remains stable for several days. We generate capped RNA from linearized plasmids using the mMessage mMachine kit (Ambion/Life technologies). However, if translation of the RNA is not of prime interest the (expensive) capping can be omitted. For nuclear injections we do not exceed a volume of 10 nL per oocyte, up to 50 nL can be injected into the cytoplasm. The RNA concentration is 1 μg/μL or less. Control oocytes are injected with an equal volume of water (*see* **Note 1**). In order to generate endo-siRNAs, transcripts with at least 30 complementary bases need to be injected into the nucleus. The resulting siRNAs are then competent to induce degradation of both nuclear and cytoplasmic reporter transcripts [8].

3 Methods

3.1 Oocyte Preparation (Defolliculation of *Xenopus laevis* Oocytes)

1. Collect parts of the ovarian lobes into a 50 mL tube half-filled with ORII solution.
2. Cut oocyte sacs into uniform, small clumps and distribute the chunks in one or two 15 mL tubes, containing 7 mL of collagenase solution (*see* **Note 2**). The final volume should not exceed 10 mL.
3. Place the tubes on a rotary mixer for approximately 1.5 h. Replace the collagenase solution and rotate for 1 h. Check regularly if follicular cells are still covering the oocytes. Stop the collagenase treatment when about 70 % of oocytes have lost the follicular layer.
4. Wash the oocytes extensively first several times with ORII, then with Barth's solution.
5. Pick healthy stage V and VI oocytes and transfer them to a petri dish with fresh Barth's solution (*see* **Note 3**).
6. Keep the oocytes overnight in Barth's solution.

3.2 Injections

1. Pull a glass capillary with a long tip. The long tip is then broken with fine forceps to generate an opening of 10–20 μM. Do not polish the tip; splintered glass cuts very easily through the oocyte membrane (*see* **Note 4**).
2. Assemble the injection apparatus according to the specific model of injector.
3. If injections into the nucleus are to be performed, set up the oocytes in the Terasaki plates and spin them at 500 × *g* for 5 min. Inject the oocyte nuclei without moving the cells from the Terasaki plate (*see* **Note 5**).
4. Inject about ten oocytes per RNA sample.
5. Transfer the oocyte to a petri dish with fresh Barth's solution and incubate at 18 °C. Check the oocytes daily and remove dead cells.
6. After the incubation that can last up to about 3–6 days the cells are ready to be processed. We have used them for protein expression, endo-siRNA production and partial reconstruction of RISC functions (Fig. 2). Downstream assays included tracer uptake measurements, western blotting, immunocytochemistry, and northern blotting. The latter is explained in detail in Chapter 8.

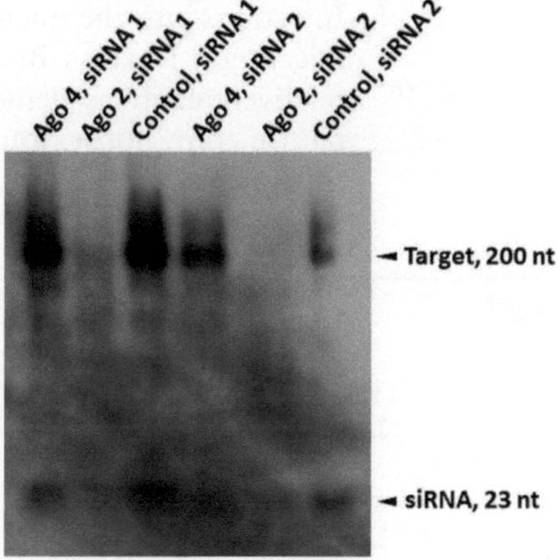

Fig. 2 Reconstitution of siRNA induced target degradation in *Xenopus laevis* oocytes. The oocytes were first injected with in vitro synthesized RNA encoding either human Argonaute 2 or Argonaute 4. The first injection took place just after sorting the oocytes (Subheading 3.1, **step 5**). The next day oocytes were reinjected with a mixture of siRNA duplex with TT overhang "oligo1" (5′-CAUCUUCACAAGGAGGAUAAGTT-3′) plus target RNA or siRNA "oligo2" (5′-GAGCAGGGAGUUGAGCAUCUUCATT-3′) plus target RNA. The target RNA consists of in vitro synthesized RNA from the oligoprobe template used for the generating the DIG labelled probes (*see* Chapter 8). The cells were incubated for 3–4 days prior to RNA extraction and northern blotting

4 Notes

1. Water injection has no effect on the oocytes, the leak seals quickly and the volume equilibrates. Therefore, we often use uninjected oocytes as a control.
2. A frog ovary contains thousands of mature oocytes suitable for experimentation; however, often far less oocytes are needed for an experiment. Depending on the number of oocytes required, either one or more aliquots should be collagenase treated.
3. Healthy oocytes have a clear separation between dark and white pole, often there is a lighter zone around the equator. Importantly, the oocyte surface should appear homogenous and not show distinct spots. The nucleus may appear as a slightly lighter area at the top of the animal (dark) pole.
4. The oocyte membrane is protected by a protein layer (vitelline) and rather resistant to damage. The opening of the tip is usually a compromise between the ease of filling the capillaries and injection and potential damage to the oocyte.

5. Injections into the nucleus can be practiced and monitored by injecting a Dextran Blue solution. The nuclei can be isolated relatively easily as demonstrated in this movie (http://www.nature.com/nprot/journal/v2/n5/extref/nprot.2007.137-S1.mov).

Acknowledgment

This work has been supported by the Dunhill Medical Trust and Newcastle University.

References

1. Okamura K, Lai EC (2008) Endogenous small interfering RNAs in animals. Nat Rev Mol Cell Biol 9:673–678
2. Chapman EJ, Carrington JC (2007) Specialization and evolution of endogenous small RNA pathways. Nat Rev Genet 8:884–896
3. Wang Q, Carmichael GG (2004) Effects of length and location on the cellular response to double-stranded RNA. Microbiol Mol Biol Rev 68:432–452
4. Tam OH, Aravin AA, Stein P, Girard A, Murchison EP, Cheloufi S, Hodges E, Anger M, Sachidanandam R, Schultz RM, Hannon GJ (2008) Pseudogene-derived small interfering RNAs regulate gene expression in mouse oocytes. Nature 453:534–538
5. Carlile M, Swan D, Jackson K, Preston-Fayers K, Ballester B, Flicek P, Werner A (2009) Strand selective generation of endo-siRNAs from the Na/phosphate transporter gene Slc34a1 in murine tissues. Nucleic Acids Res 37:2274–2282
6. Watanabe T, Totoki Y, Toyoda A, Kaneda M, Kuramochi-Miyagawa S, Obata Y, Chiba H, Kohara Y, Kono T, Nakano T, Surani MA, Sakaki Y, Sasaki H (2008) Endogenous siRNAs from naturally formed dsRNAs regulate transcripts in mouse oocytes. Nature 453:539–543
7. Soumillon M, Necsulea A, Weier M, Brawand D, Zhang X, Gu H, Barthes P, Kokkinaki M, Nef S, Gnirke A, Dym M, de Massy B, Mikkelsen TS, Kaessmann H (2013) Cellular source and mechanisms of high transcriptome complexity in the Mammalian testis. Cell Rep 3:2179–2190
8. Carlile M, Nalbant P, Preston-Fayers K, McHaffie GS, Werner A (2008) Processing of naturally occurring sense/antisense transcripts of the vertebrate Slc34a gene into short RNAs. Physiol Genomics 34:95–100
9. Saccomanno L, Bass BL (1994) The cytoplasm of Xenopus oocytes contains a factor that protects double-stranded RNA from adenosine-to-inosine modification. Mol Cell Biol 14:5425–5432
10. Markovich D (2008) Expression cloning and radiotracer uptakes in Xenopus laevis oocytes. Nat Protoc 3:1975–1980
11. Tammaro P, Shimomura K, Proks P (2008) Xenopus oocytes as a heterologous expression system for studying ion channels with the patch-clamp technique. Methods Mol Biol 491:127–139

Chapter 4

Analysis of Endo-siRNAs in Drosophila

Katharina Elmer, Stephanie Helfer, Milijana Mirkovic-Hösle, and Klaus Förstemann

Abstract

The small RNA silencing system is an important player in the control of gene expression. In particular analysis of the highly complex sequence repertoire of piRNAs and endogenous siRNAs directed against repetitive elements and transposons has been facilitated by the advent of next-generation sequencing tools. We are providing a detailed protocol for the creation of deep sequencing libraries in combination with a chemical modification step (periodate oxidation and β-elimination) that enriches for 3′-terminal modified small RNAs. This step can thus facilitate the detection and—since incorporation of the small RNA into the effector complex precedes the terminal methylation—determine the Argonaute-loading state of certain small RNA species. Both, the oxidation as well as the deep sequencing library preparation are adaptations of popular, previously published protocols. The sequencing libraries described here can be sequenced on all flow cell types (i.e., single-end and paired-end) of the Illumina sequencing systems and are compatible with Illumina index reads. It is therefore also possible to analyze the libraries on the MiSeq system, for which currently only paired-end flow cells are available, and combine several libraries in multiplexed experiments. The chapter concludes with considerations concerning quality control and data analysis.

Key words Deep sequencing, Next-generation sequencing, Illumina TrueSeq Index, MiSeq system, miRNA, Endo-siRNA, esiRNA, Endogenous siRNA

1 Introduction

Endogenous siRNAs can be defined either by their origin (mapping to the host genome) or by their biogenesis route, which closely follows the one established for siRNAs, rather than the one of miRNAs [1–4]. Depending on the organism, this may lead to a more or less precisely defined set of small RNAs. For example in *Drosophila*, this definition would include two rather distinct groups of small RNAs, the long hairpin-derived small RNAs and endo-siRNAs from RNA hybrids from complementary transcripts. The former are defined as products from long hairpin-forming transcripts with longer stems than miRNA precursors. The hairpin-derived endo-siRNAs are of low sequence complexity and individual sequences are found at high abundance, comparable to miRNAs.

Andreas Werner (ed.), *Animal Endo-siRNAs: Methods and Protocols*, Methods in Molecular Biology, vol. 1173,
DOI 10.1007/978-1-4939-0931-5_4, © Springer Science+Business Media New York 2014

Messenger RNA targets for these endo-siRNAs can be predicted with algorithms similar to those used for miRNA target predictions. Accordingly, endo-siRNA-directed target cleavage has been demonstrated but mRNA destabilization and/or translational inhibition may also be invoked. Thus, the biogenesis of these endo-siRNAs resembles siRNAs whereas their mode of action is more akin to miRNAs. On the other hand, endo-siRNAs from complementary transcripts represent a complex mixture of small RNAs where each individual sequence is of low abundance. Under special circumstances it may be possible to detect some of these siRNAs by low-sensitivity techniques like northern blotting; however, a representative view can only be obtained through deep sequencing. This technique was also at the origin of endo-siRNA discovery and mapping of the reads pointed to a biological role of these endo-siRNAs: The repression of transcripts from transposable element in somatic and germ line cells. Comparable approaches have since identified other roles, such as chromatin modification [5–8], alternative splicing [9, 10], and recently DNA double-strand break repair [11–13]. Next-generation sequencing is thus the method of choice for the analysis of endo-siRNAs.

In *Drosophila* the biogenesis of Argonaute 2-loaded (Ago2) small RNAs includes an additional modification step: The terminal 2′-hydroxyl-group is modified with a 2′-*O*-methyl group by the methyl-transferase DmHen-1. This modification is fairly common among small RNAs; in animals it can be found in piRNAs (in *Drosophila* loaded onto Piwi, Aubergine, and Ago3) and also in many small RNAs found in plants. The purpose of this modification is likely to protect the small RNA guide from nucleotidyl transferase activity that can induce the degradation of the small RNA after binding to a perfectly base-pairing target RNA. This mechanism selects against (unmodified) miRNAs with perfect complementary to off-targets, whereas piRNAs and siRNAs which are intended to function via perfect complementarity are modified. However, human Ago2-loaded small RNAs which are not 2′-*O*-methyl modified still silence fully complementary targets efficiently indicating that the terminal transferase activity is low and allows for some target RNA degradation.

In the following sections we describe an experimental procedure that enriches small RNAs carrying the terminal 2′-*O*-methyl modification prior to library generation. Such a strategy reduces the presence of miRNAs and concomitantly increases the sequencing depth for endo-siRNAs. Since Ago-loading precedes 2′-*O*-methyl modification, this approach can also be exploited to demonstrate loading of a given set of small RNAs into the respective Ago-protein. The approach and its implementation for deep sequencing have been pioneered by Phil Zamore and his collaborators [14–17]; here we present a slightly modified version of the protocol. Importantly, it adapts a popular small RNA library generation protocol [18] to enable sequencing on Illumina paired-end design flow cells, a prerequisite for sequencing on the MiSeq System. The procedures have been optimized for *Drosophila* RNA but will be applicable to RNAs from any organism or tissue source, provided the 2′-*O*-methyl modification is present.

For reasons of clarity, we refrain from describing certain procedures in detail (e.g., denaturing acrylamide/urea gel electrophoresis of nucleic acids) since standard protocols will suffice.

2 Oxidation and β-Elimination of RNAs

2.1 Material and Solutions

1. 5× Borate incubation buffer:

   148 mM borax.

   148 mM boric acid, pH 8.6.

   2.5 % SDS (can be prepared in advance, for example as 10 % w/v stock solution).
2. *$NaIO_4$ solution*: freshly prepare a 200 mM solution of $NaIO_4$ in RNase-free water.
3. *NaOH solution*: 2 M (we buy a ready-made solution and keep it RNase-free).
4. *Glycogen solution*: 20 μg/μL (Roche Molecular Diagnostics).

Microcentrifuge-based gel filtration columns: Removal of salts and other small molecules can be done efficiently using small gel filtration columns and a microcentrifuge. We recommend a commercial source (e.g., Roche Mini Quick-Spin Oligo) as the columns must be RNase free.

2.2 Oxidation of Vicinal Hydroxyl-Groups

1. Combine the following reaction in a microcentrifuge tube:

20.25 μL	of total RNA dissolved in water, the recommend amount is 10–20 μg
6 μL	of 5× borate incubation buffer (148 mM borax, 148 mM boric acid pH 8,6, 2.5 % SDS)
3.75 μL	of $NaIO_4$ solution (200 mM) (*see* **Note 1**)

   Incubate the mixture for 30 min at room temperature. Then add 3 μL of 100 % glycerol to quench the remaining $NaIO_4$ and incubate for 10 min at room temperature.
2. β-elimination in alkaline pH: Add 2.5 μL of 2 M NaOH solution to the reaction mixture and incubate for 90 min at 45 °C (*see* **Note 2**).
3. Purification of modified RNA: Purification of the modified RNA is necessary at this point (with- or without β-elimination) because the SDS in the sample leads to an insoluble precipitate

upon the addition of ethanol. Prepare the column according to the *manufacturer's instructions* and apply the entire reaction mixture onto the column. Centrifuge and collect the eluate.

4. Add 1 µL of glycogen solution (20 µg/µL, this serves as a carrier for precipitation) and approximately 200 µL of 100 % ethanol (200 µL corresponds to three volumes of the column eluate, adjust if necessary).

Mix thoroughly, then centrifuge for 20 min at maximum speed in a microcentrifuge, discard the supernatant. Wash the pellet twice with 150 µL of 70 % ethanol. *The second washing step should be left for at least 30 min, ideally at −20°C overnight, to remove any remaining salt prior to centrifugation.* Remove excess liquid, dry for 5 min at room temperature (the lid can be kept closed to prevent accidental introduction of traces of RNase) and dissolve the pellet in 10 µL of water (*see* **Note 3**).

Use 1 µL of the final product for quality control on a 20 % denaturing acrylamide/urea gel (a mini-gel setup as used for Western Blots will do just fine). After staining with, e.g., Sybr gold, the RNA bands can be visualized and documented. In the case of *Drosophila* RNA, shortening of the 30 nt long 2S rRNA can be easily seen if an untreated and a beta-eliminated sample are loaded side by side. All RNA bands should be sharp (no smears!) and the 2S rRNA should be completely shifted to the smaller size. An example is shown in Fig. 1a.

The RNA is now ready for library generation according to the protocol outlined in part 4.

3 Generation of Small RNA Libraries for Sequencing on the Illumina Platform

We isolate total RNA using Trizol or a related reagent. The quality of this RNA is usually very good and the method returns high yields. There are no further purification steps required prior to oxidation and elimination; however, a substantial amount of RNA should be used at start since the precipitation and purification steps included in this protocol will lead to loss of material.

Because the barcode is introduced only at the final step, special care must be taken to avoid cross-contamination of samples! Ideally, a separate bench with a dedicated set of pipettes should be set up, and filter tips should be used for all steps. Keep in mind that this is a PCR procedure; under other circumstances contaminations may fall below the detection limit of agarose gel electrophoresis, the sensitivity of deep sequencing will be much less forgiving. *During all gel-purification steps, only one sample should be run per gel and gel chamber* (= do not run two different samples in the same chamber back-to-back on different gels).

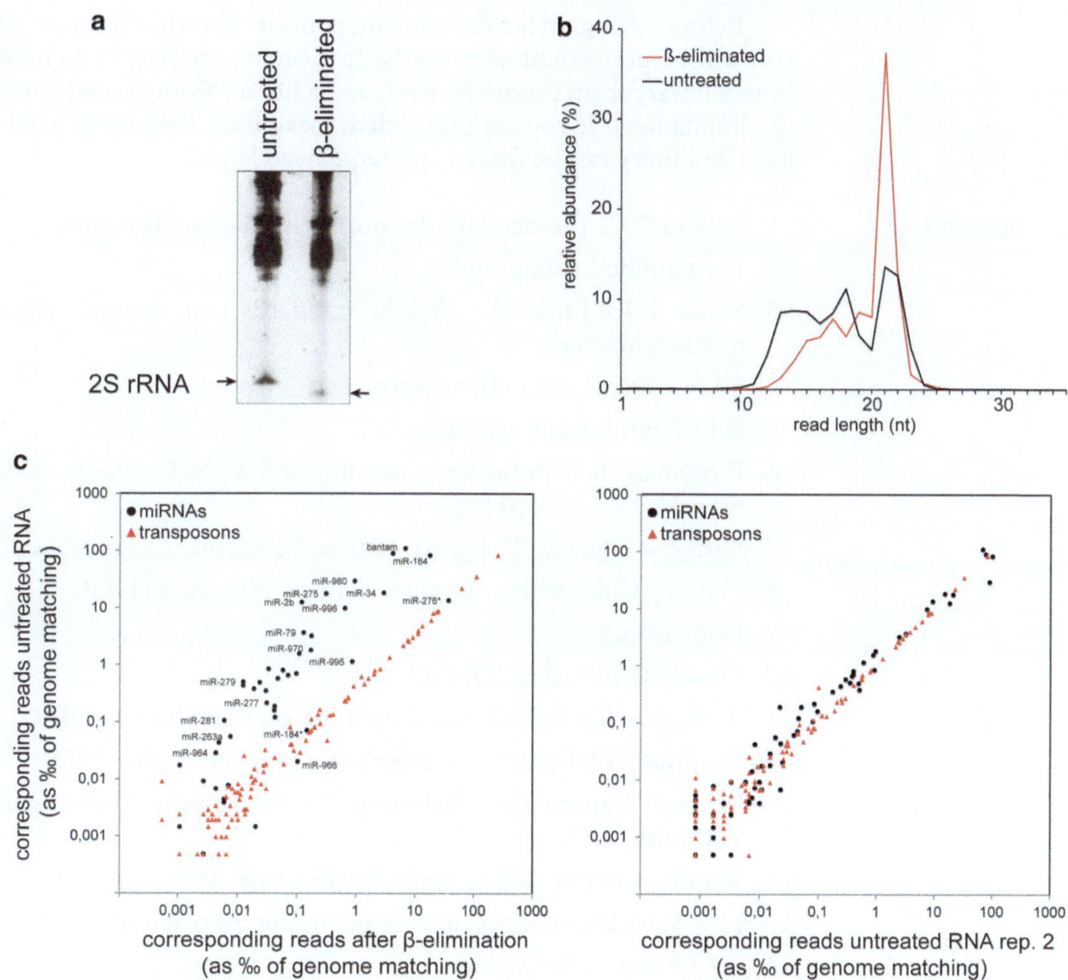

Fig. 1 Analysis of β-eliminated RNA: (**a**) RNA sample before and after β-elimination, run on a 20 % acrylamide/urea mini-gel and stain with Sybr gold. The 30 nt long 2S rRNA is indicated; its size decreases by one nucleotide upon oxidation and beta-elimination. A double band indicates inefficient oxidation and/or elimination reactions. In that case, increase the incubation times at either step until the gel image shows a complete shift of the 2S rRNA band. (**b**) Upon β-elimination, a clear change in the size distribution profile of the sequenced reads should be visible. The presented example was obtained with RNA from *Drosophila* S2 cells. After the treatment, the 21 nt peak is clearly enhanced in this example and the number of miRNA reads (with a predominant length of 22 nt) is reduced. If, for example, ovarian RNA was analyzed, a prominent piRNA peak in the range of 23–28 nt, corresponding to piRNAs should also be visible and enhanced upon treatment. Some degradation of the sample may have occurred during incubation at high pH for β-elimination resulting in the shoulder in the ~14–20 nt range. If this degradation should be a concern the RNA may only be oxidized but not β-eliminated prior to library generation. Alternatively, one can restrict the analysis to 21 nt or longer reads. (**c**) Two-dimensional plots of reads corresponding to individual miRNAs or transposons can reveal that transposon-matching endo-siRNAs are not affected by the chemical treatment, while miRNA reads are substantially depleted. Within the group of miRNAs, clear differences can nonetheless be seen. Ago-sorting rules in *Drosophila* state that the star-strand or passenger of miRNAs is more likely to be incorporated into Ago2 and, accordingly, miR-276* as well as miR-184* clearly behave more like endo-siRNAs. Another known example is miR-277, a miRNA that is 30–50 % Ago2-loaded belonging to both the miRNA and endo-siRNA groups. For clarity, only selected miRNA data points are identified with a specific label. The diagram on the *right* shows the reproducibility of read quantification between two independent biological replicates using untreated RNA. The changes observed upon β-elimination should clearly exceed biological and technical variability

Before setting up the experiment, consider also the data analysis you will be performing afterwards. It is often very helpful to have both a library from untreated RNA and a library from treated RNA (β-elimination). If you are undecided, make both libraries in parallel. The library can be frozen and sequenced later.

3.1 Materials

1. 15 and 20 % polyacrylamide/urea gels (Sequagel system).
2. Formamide loading buffer.
3. Small RNA Molecular Weight standards (for example NEB miRNA marker).
4. 50 bp DNA ladder (Invitrogen).
5. Sybr Gold nucleic acid stain.
6. Proteinase K with buffer containing 0.4 M NaCl, 0.5 % SDS, 50 mM Tris–HCl pH 8.
7. *Glycogen solution*: 20 μg/μL (Roche Molecular Diagnostics).
8. Phenol/Chloroform/Isoamyl alcohol mixture, pH 8.0.
9. Isopropanol.
10. Dimethyl sulfoxide (DMSO).
11. *Truncated* T4 RNA ligase 2 with the supplied buffer (NEB).
12. (Normal) T4 RNA ligase 1 with the supplied buffer (NEB).
13. Reverse transcriptase including 5× first strand buffer and 100 mM DTT.
14. RNase inhibitor (for example RNasin from Promega).
15. Hot Start Phusion polymerase including 5× reaction buffer.
16. dNTP mix (10 mM each).
17. Gel Extraction Kit (Qiagen).
18. Required oligonucleotides:

First ligation at the 3′-end:

Linker 1	AMP-5′p = 5′pCTGTAGGCACCATCAATdideoxyC-3′
	Pre-adenylated linker, commercially available from IDT
	Product name: miRNA cloning linker 1
	Prepare aliquots of 50 μM stock solution

Second ligation at the 5′-end:

Illumina linker	5′-rArCrArCrUrCrUrUrUrCrCrCrUrArCrArCrGrArCrGrC rUrCrUrUrCrCrGrArUrCrU-3′
	Our oligonucleotide supplier is MWG/Eurofins
	Quality: HPLC purified, prepare aliquots of 50 μM stock solution

Reverse transcription:

RT primer index	5′-GTGACTGGAGTTCAGACGTGTGCTCTTCCGATC
	GATTGATGGTGCCTACAG-3′
	MWG—Custom oligo—HPSF purified, 5 μM stock
	(*underlined: Illumina index primer binding site*)

PCR amplification:

5′-Illumina paired	5′-AATGATACGGCGACCACCGAGATCTACACTCTTTCCCT ACACGACG-3′ (*underlined sequence: necessary for compatibility with paired-end flow cells on the Illumina systems*) MWG—Custom oligo—HPSF purified, aliquots of 10 μM stock
3′-PCR index1	5′-CAAGCAGAAGACGGCATACGAGATatcacgGTGACTGGAGTTCA GACGTG -3′
3′-PCR index2	5′-CAAGCAGAAGACGGCATACGAGATcgatgtGTGACTGGAG TTCAGACGTG -3′
3′-PCR index3	5′-CAAGCAGAAGACGGCATACGAGATttaggcGTGACTGGAG TTCAGACGTG -3′
3′-PCR index4	5′-CAAGCAGAAGACGGCATACGAGATtgaccaGTGACTGGAG TTCAGACGTG -3′
3′-PCR index5	5′-CAAGCAGAAGACGGCATACGAGATcagaatGTGACTGGAG TTCAGACGTG -3′
3′-PCR index6	5′-CAAGCAGAAGACGGCATACGAGATgctgtaGTGACTGGAG TTCAGACGTG -3′
3′-PCR index7	5′-CAAGCAGAAGACGGCATACGAGATtcgcacGTGACTGGAG TTCAGACGTG -3′
3′-PCR index8	5′-CAAGCAGAAGACGGCATACGAGATcagtggGTGACTGGAG TTCAGACGTG -3′
3′-PCR index9	5′-CAAGCAGAAGACGGCATACGAGATggtatcGTGACTGGAG TTCAGACGTG -3′
	All index primers: MWG/Eurofins, HPSF purified, aliquots of 10 μM stock (*underlined sequence: necessary for compatibility with paired-end flow cells on the Illumina systems*)

4 Methods

1. Selection of small RNAs by denaturing gel electrophoresis.

Prepare a 20 % polyacrylamide/urea gel (Sequagel system, National Diagnostics; 8 mL concentrate, 1 mL diluent, 1 mL buffer) or use a commercially available precast gel system; pre-run gel for 5 min at 300 V prior to loading.

Heat RNA sample mixed with at least an equal volume of formamide containing gel loading buffer to 70 °C for 5 min (we recommend using 10–20 μg of total RNA).

Load 2 μL of formamide loading buffer in each well; this identifies bad wells and makes the gel run more even, then load 5 μL miRNA marker for size control (NEB, three bands: 17, 21, and 25 nt).

Load the sample. Run the gel at constant 250 V for 45–60 min.

Stain gel 5 min with Sybr Gold in a large Petri dish (5 μL SyBr Gold in 50 mL of water).

On a transilluminator, excise the region corresponding to the desired size of small RNAs (~16–28 nt) and place it into a 0.5 mL microcentrifuge tube. It is convenient to use the 2S rRNA (30 nt) band for orientation since in most cases the si/miRNA fraction will be invisible.

Pierce the 0.5 mL tube at the bottom three times with a .22 G needle, insert it into a 1.5 mL microcentrifuge tube, and spin for 2 min at full speed to shred gel slice.

Add 500 μL of proteinase K buffer (0.4 M NaCl, 0.5 % SDS, 50 mM Tris–HCl pH 8) and then 1 μL of proteinase K.

Shake for *at least* 2 h at 65 °C in a thermomixer to elute RNA, invert tube from time to time. Alternatively, rotate at room temperature over night.

To remove the gel remnants, spin the gel slice homogenate through an empty spin column (2 min at full speed).

Add 20 μg of glycogen (1 μL of the stock solution), extract solution with 400 μL of the phenol/chloroform/isoamyl alcohol mixture (pH 8.0), and centrifuge for 20 min at maximum speed in a microcentrifuge.

Transfer the aqueous supernatant to fresh tube and precipitate with 400 μl of isopropanol: mix well, incubate for 15 min at room temperature, spin for 20 min full speed at room temperature, discard the supernatant.

Wash pellet 2× with 150 μL of 70 % ethanol, let second wash stand for at least 30 min, spin as above, and discard the supernatant carefully.

Leave tube with lid *closed* for 1 min to dry RNA. Resuspend the RNA in 8 μl of water.

2. Linker ligation to the 3′-end of small RNAs.
   Assemble in a microcentrifuge tube the following reaction:

6 μL	Size-selected RNA dissolved in water
1 μL	ATP-free T4 RNA ligase buffer (10×, supplied with ligase)
1 μL	DMSO
1 μL	Oligo Linker 1 (50 μM stock)
1 μL	Truncated T4 RNA ligase 2

   Incubate the reaction for 30 min at 25 °C, then heat-inactivate the ligase for 5 min at 95 °C. At this point, the remaining, unreacted Linker 1 must be removed by gel purification.

3. Purification of small RNAs by denaturing gel electrophoresis. This step occurs essentially as described for step B1, except that a 15 % acrylamide-urea gel should be used. In addition, because of the increased size of the small RNAs due to linker ligation the 50 bp DNA ladder should be used as marker.

   Excise a gel slice that places the 50 nt DNA band within its upper third and, at its lower end, is well separated from the unreacted Linker 1. You may load an aliquot of the Linker 1 oligonucleotide in a neighboring well for clarity.

4. Adapter ligation to the 5′-end of small RNAs.
   Assemble in a microcentrifuge tube the following reaction:

6 μL	RNA, gel-purified after 3′ linker ligation, dissolved in water
1 μL	T4 RNA ligase buffer (10×, supplied with ligase) with ATP
1 μL	DMSO
1 μL	Oligo *Illumina Linker* (50 μM stock)
1 μL	Normal T4 RNA ligase 1

   Incubate the reaction for 60 min at 25 °C, then heat-inactivate the ligase for 5 min at 95 °C (*see* **Note 4**).

5. Reverse transcription.
   Assemble in a microcentrifuge tube the following reaction:

9 μL	5′-adapter ligation reaction from **step 4**
2 μL	Oligo RT primer index (5 μM)

   Denature at 95 °C for 2 min, then cool the reaction on ice for 2 min. The add:

4 μL	5× first strand buffer (supplied with the reverse transcriptase)
2 μL	100 mM DTT (supplied with the enzyme)
1 μL	RNase inhibitor (e.g., RNasin from Promega)
1 μL	dNTP mix (10 mM each)

Mix, then split the reaction into two tubes of 9 μL each. Incubate for 2 min at 42 °C. Then add either

1 μL	Superscript II RT enzyme (Invitrogen, +RT sample)
or	
1 μL	H_2O (-RT sample)

Incubate the reactions at 42 °C for 30 min, then heat-inactivate the enzyme for 5 min at 95 °C.

Continue directly with **step 6** or store the samples at –20 °C until use.

6. PCR amplification of cDNA.
At this step, sequence barcodes for multiplexed library sequencing are introduced. It is important to consider how many samples are going to be pooled and choose the indexes accordingly. Since the barcode sequences will be of low complexity, a barcode combination should be chosen that will have at least one A or C and one G or T at each position (*see* legend to Fig. 2). Especially if only two libraries are combined (for example a small-scale experiment run on a MiSeq system) de-multiplexing may prove challenging if a unsuitable index combination was used. We indicate suitable combinations of the barcodes described in this protocol in Fig. 2.

Assemble in two PCR tubes the following reactions:

5 μL	First strand cDNA (+RT) or control (–RT) from **step 5**
20 μL	5× PCR buffer (including Mg^{2+}, 2.5 mM final concentration)
2 μL	dNTP mix (10 mM each)
1 μL	oligo *5′ Illumina paired end* (10 μM)
1 μL	oligo *3′ PCR index* 1–6 (10 μM, see above for index combinations)
70 μL	H_2O
1 μL	Hot Start Phusion polymerase

Run the following thermocycler program:

98 °C	1 min	(Initial denaturation and enzyme activation)
98 °C	15 s	Denaturation
58 °C	30 s	Annealing
72 °C	30 s	Extension
72 °C	2 min	Final extension
4 °C	pause	

(*see* **Notes 5** and **6**).

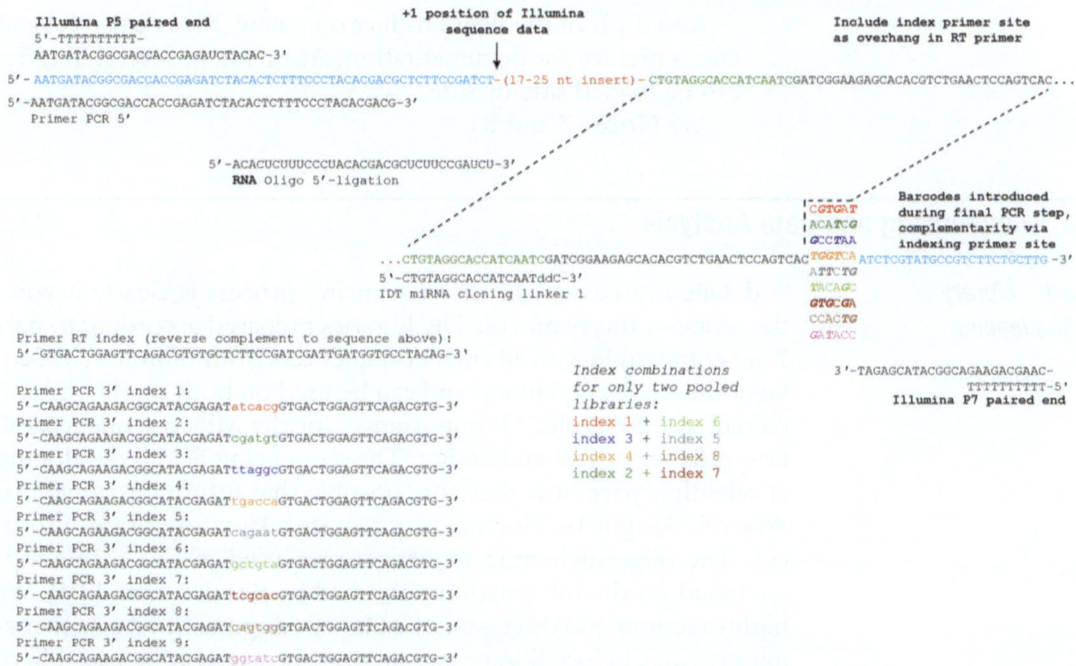

Fig. 2 Sequence scheme of the final library design. For simplicity only the top strand of the final product is indicated, displayed in the 5′–3′ direction. The sequence is represented on *two lines* and in part duplicated for clarity; the dashed lines indicate the duplicated portion in the *lower line*. The position of the indexes is specified by the multiple sequence alternatives and the indexes are *color-coded*. Corresponding primer sequences can be found on the *left*, always written in 5′–3′ direction of the primer. Note that the index portion in the primer sequence on the *bottom left* is the reverse complement of the index read. The Illumina Sequencers need signal in both image channels (*red* and *green* excitation or GT and AC) at every cycle. For a mixture of only two distinct indexes, combinations must be chosen that ensure this is true. With the index sequences we are proposing this is true for index combinations 1 + 6 or 3 + 5. These combinations should thus be the basis for the addition of further indexes

7. Gel purification of the PCR products.

   This step follows standard molecular biology practice. Prepare a 2 % agarose gel with a large well comb. The 50 bp ladder should be used as size marker. Depending on your setup, you may have to distribute the PCR into several pockets. To avoid cross-contaminations do not run different libraries in parallel on the same gel (*see* above).

   Excise the band at around 160 nucleotides; make sure a potentially visible, slightly smaller band is not included. This approximately 20 bp smaller band corresponds to amplified linker-to-linker ligation product. Purify the excised DNA using a gel extraction kit (use 1 column per gel slice if the sample was run in two pockets). If you used two columns, elute the sample sequentially using the same buffer for both columns (we recommend using 30 μL). A scheme of the final library ready for sequencing is included in Fig. 2.

Run 3 μL of the final product on a new 2 % agarose gel and take a picture for documentation. At this step, several libraries can be loaded side by side.

(*see* **Notes 7** and **8**)

5 Sequencing and Data Analysis

5.1 Library Sequencing

A detailed discussion of the sequencing process is clearly beyond the scope of this protocol. The libraries prepared according to part 2 are compatible with all current sequencers from Illumina (MiSeq, Genome Analyzer, HiSeq) and can be used on both single-end and paired-end flow cells. Of importance, for the MiSeq sequencer, all flow cells are paired-end design. The sequencing field is developing at relentless pace; it is therefore possible that future instrument or flow cell designs by Illumina may require adaptation of this protocol. The oligonucleotide sequences presented in this manuscript are based on the information published by Bentley et al. [19]. We highly recommend discussing the library design with your sequencing provider before library preparation. It may also be possible to change the oligonucleotide sequences in order to use this protocol on other sequencing platforms. Please refer to Fig. 2 for an annotation of the functional elements within our library design.

The index sequences we are proposing here are *not* the original Illumina indexes. You can in principle combine libraries made according to this protocol with other libraries made according to the Illumina protocol (e.g., for a HiSeq 2500 run, double-check the indexes before library generation!); however, we have not tested this. It is certainly possible to exchange our indexes with the Illumina index sequences in the PCR primers (they can be obtained from your sequencing provider or found on the Illumina support pages). In this case you may directly use the sample sheet creator application and other software pipelines from the Illumina suite. Alternatively, it is possible to edit the index sequence within the sample sheet manually after creating it with arbitrary indexes using a text editor; this strategy will also result in direct de-multiplexing on the MiSeq system.

The Illumina Genome Analyzer GAIIx and HiSeq system are usually operated by sequencing facilities; we therefore encourage you to contact your provider for more information. Since the MiSeq system is often more directly accessible we will provide some more information below. Common to all platforms is the need for a precise quantification of the final library (or mix of libraries). This is to ensure optimal cluster density for high-quality, high-yield sequencing runs. Common options for quantification are either qPCR or a Bioanalyzer run. For qPCR you will need a reference sample for calibration; the Bioanalyzer uses internal calibration standards. In our experience the libraries prepared according to this protocol can show abnormal migration behavior. In this case, the number of PCR cycles for amplification should be reduced and/or the amount of polymerase increased.

The following options should be chosen on the MiSeq system:

Library type:	Illumina TrueSeq LT
Workflow:	Generate Fastq
	The 3′-Linker sequence must be trimmed prior to further analysis; as it is not the same sequence as in the original TrueSeq libraries, this step has to be performed separately post-run.
Read 1:	Single-end, e.g., 1 × 35 nt
	This is significantly longer than required for the analysis of mi/siRNAs; however, the data will include small amounts of 2S rRNA reads, which can be used for quality control and/or normalization purposes. If you focus exclusively on miRNAs and siRNAs, 1 × 25 nt will be enough and you save running time.
Index read 1:	6 nt

The output will be a set of fastq files that are already de-multiplexed and numbered according to the index. There are always two corresponding files, one containing the sequence reads in one file and the index reads in the other file. Any indexes that cannot be interpreted and the corresponding sequence reads are collected in an additional file pair for unassigned reads. If any mistakes occurred during sample sheet editing, this file pair can subsequently be reinterpreted and de-multiplexed offline.

We are happy to provide short PERL scripts that can be used for offline de-multiplexing, trimming of the 3′-linker sequence, and size-selection of the obtained small RNA sequences. It is usually advisable to exclude reads shorter than 16 nt from analysis as these may be difficult to be mapped unambiguously. The scripts can be run in a terminal window on a Linux operating systems (we do not recommend using a Windows PC due to file size limitations). Please contact the corresponding author with your request.

5.2 Mapping and Data Analysis

Mapping of the reads can be performed with any of the available software tools, we routinely use bowtie [20]. In the case of transposon-repressing endo-siRNAs, however, the choice of the reference sequence is important. One possibility is to map to a reference genome, then use the annotation of transposon or repeat sequences to assign the small RNA reads to these targets. A different approach is to use a species-specific collection of transposon sequences as the "genome" to map to. In the first case, small sequence variations among multiple copies of the same transposon inserted at different positions may make it possible to unambiguously assign some reads to a genomic site of origin. Nonetheless when studying active transposons, which are usually present in multiple copies per genome, it is sometimes inappropriate to limit the analysis to only those reads that map uniquely within the genome—they may simply not be representative of the bulk. It should also be taken into account that we rarely use strains or even cell cultures that are derived from the sequenced reference strain. In fact, since transposable elements are by definition mobile, fly lines that correspond to the reference strain including the location of active transposons likely do not exist anymore. A uniquely mapping, transposon matching read could therefore very well be a simple coincidence—a conundrum that is impossible to solve. If the reads are mapped to compiled transposon sequences their genomic site of origin cannot be identified. In that case it is *essential* to limit the analysis to reads that match perfectly to the reference sequence (i.e., with no mismatch allowed)—this ensures that there is no other position in the *Drosophila* genome where the read might fit even better. This approach facilitates downstream analyses, but has the disadvantage that small sequence variations from the canonical sequence present in the collection used to define the "genome" will be excluded. Which of the mapping strategies is most appropriate depends on the particular question that you are pursuing with the deep sequencing experiment.

Data analysis is specific to each project. We will therefore limit our discussion here to the aspect of quality control of the oxidation/β-elimination reaction. It is important to verify that the chemical treatment has been exhaustive. In addition to gel electrophoresis and staining of the β-eliminated RNA (*see* **Note 3**), we propose to first analyze the general size distribution of the small RNA reads obtained. If possible, you can compare the size distribution of untreated and treated libraries as demonstrated in Fig. 1b. The corresponding size class (e.g., 21 nt length for *Drosophila* endo-siRNAs) should be clearly increased after oxidation/β-elimination and other read lengths decreased. If you do not observe any changes it is likely that the chemical treatment did not work. In a second step, a two-dimensional plot of normalized read numbers corresponding to individual miRNAs and transposons in libraries derived from treated and untreated RNA can be generated (*see* Fig. 1c). In this plot, the miRNAs should be clearly separated from the transposons. Furthermore, since certain miRNAs are partially Ago2-loaded, the miRNA data points should have a wider spread than the data points corresponding to transposons. For comparison, we included a corresponding plot comparing independent biological replicates of untreated RNA.

6 Notes

1. It is essential to prepare the $NaIO_4$ solution freshly using crystalline $NaIO_4$!
2. In principle, β-elimination is not required for the generation of deep sequencing libraries; the incubation at high temperature and alkaline pH bear the risk of RNA degradation.

However, verification of the oxidation reaction is only possible after elimination via gel electrophoresis and staining of the RNA (β-eliminated 2S rRNA shows enhanced mobility). If you have enough material, you can consider splitting the sample and use one part for cloning and the other for β-elimination.

3. Use 1 μL of the final product for quality control on a 20 % denaturing acrylamide/urea gel (a mini-gel setup as used for western blots will do just fine). After staining with Sybr gold, the RNA bands can be visualized and documented. In *Drosophila* RNA, shortening of the 30 nt long 2S rRNA can be easily documented if an untreated and a β-eliminated sample are loaded side by side. All RNA bands should be sharp (no smears!) and the 2S rRNA should be completely shifted to the smaller size. An example is shown in Fig. 1a.

4. In our hands, a gel-purification step is not required prior to reverse transcription. However, if you find high levels of empty products in your libraries (linker-to-linker ligations) an additional gel-purification step can help. Alternatively, increasing the amount of starting material will also reduce the extent of linker-to-linker ligations.

5. 15–25 cycles may be required; determine empirically how many cycles are needed until a band is clearly visible in an agarose gel stained with Sybr Safe or ethidium bromide. Do not over-amplify!

6. The specific brand of Taq polymerase is not important. However, we recommend using a hot-start enzyme with high fidelity since nonspecific amplification products may adversely affect sequencing and polymerase mistakes can interfere with adapter trimming and read mapping during data analysis.

7. You can clone the PCR products with the help of a standard blunt-end cloning kit (for example the pJet system from Fermentas) and sequence the inserts from 5 to 10 colonies by standard Sanger sequencing for quality control purposes. This will give a reasonable estimate of the number of "empty" ligation products (linker-to-linker ligation) and can also indicate whether the flanking sequences have been incorporated as expected. If recurrent sequence errors are found, poor oligonucleotide quality may be the cause. In that case, reorder the corresponding oligonucleotide and include HPLC or PAGE purification.

8. In our experience, the relative concentrations of each library can be estimated after running aliquots of the libraries side by side on a gel. This information can be sufficient to determine the mixing ratio (but not the final concentration!) for multiplexed libraries. Alternatively, the concentration of individual or multiplexed libraries can be determined by qPCR or a Bioanalyzer.

9. Let's assume that the oxidation reaction was 99 % complete (a very respectable value for chemical derivatization); a maximal shift of two orders of magnitude is expected in the read numbers upon treatment for an miRNA that is exclusively loaded into Ago1. In the example presented in Fig. 1c, the strongest depletion seen for individual miRNAs is 97–99 %. Determining whether the remaining 1–3 % of RNAs were Ago1-loaded but escaped treatment or were Ago2-loaded (and thus modified) would require further analysis. The maximal achievable level of depletion will depend on the origin and the nature of your RNA samples; our example was RNA obtained from a clonal population of S2 cells.

Acknowledgements

We would like to thank Romy Böttcher for help with experiments to validate the deep sequencing protocol.

References

1. Ghildiyal M, Seitz H, Horwich MD, Li C, Du T, Lee S, Xu J, Kittler EL, Zapp ML, Weng Z, Zamore PD (2008) Endogenous siRNAs derived from transposons and mRNAs in Drosophila somatic cells. Science 320:1077–1081
2. Czech B, Malone CD, Zhou R, Stark A, Schlingeheyde C, Dus M, Perrimon N, Kellis M, Wohlschlegel JA, Sachidanandam R, Hannon GJ, Brennecke J (2008) An endogenous small interfering RNA pathway in Drosophila. Nature 453:798–802
3. Kawamura Y, Saito K, Kin T, Ono Y, Asai K, Sunohara T, Okada TN, Siomi MC, Siomi H (2008) Drosophila endogenous small RNAs bind to Argonaute 2 in somatic cells. Nature 453:793–797
4. Okamura K, Chung WJ, Ruby JG, Guo H, Bartel DP, Lai EC (2008) The Drosophila hairpin RNA pathway generates endogenous short interfering RNAs. Nature 453:803–806
5. Cernilogar FM, Onorati MC, Kothe GO, Burroughs AM, Parsi KM, Breiling A, Lo Sardo F, Saxena A, Miyoshi K, Siomi H, Siomi MC, Carninci P, Gilmour DS, Corona DF, Orlando V (2011) Chromatin-associated RNA interference components contribute to transcriptional regulation in Drosophila. Nature 480:391–395
6. Verdel A, Jia S, Gerber S, Sugiyama T, Gygi S, Grewal SI, Moazed D (2004) RNAi-mediated targeting of heterochromatin by the RITS complex. Science 303:672–676
7. Zofall M, Grewal SI (2006) RNAi-mediated heterochromatin assembly in fission yeast. Cold Spring Harb Symp Quant Biol 71: 487–496
8. Grewal SI, Elgin SC (2007) Transcription and RNA interference in the formation of heterochromatin. Nature 447:399–406
9. Harel-Bellan A, Zazoua MA, Rachez C, Muchardt C, Batsche E (2013) 10-million-years AGO: argonaute on chromatin in yeast and human, a conserved mode of action? Transcription 4:89–91
10. Ameyar-Zazoua M, Rachez C, Souidi M, Robin P, Fritsch L, Young R, Morozova N, Fenouil R, Descostes N, Andrau JC, Mathieu J, Hamiche A, Ait-Si-Ali S, Muchardt C, Batsche E, Harel-Bellan A (2012) Argonaute proteins couple chromatin silencing to alternative splicing. Nat Struct Mol Biol 19:998–1004
11. Michalik KM, Bottcher R, Forstemann K (2012) A small RNA response at DNA ends in Drosophila. Nucleic Acids Res 40:9596–9603
12. Wei W, Ba Z, Gao M, Wu Y, Ma Y, Amiard S, White CI, Danielsen JM, Yang YG, Qi Y (2012) A role for small RNAs in DNA double-strand break repair. Cell 149:101–112
13. Francia S, Michelini F, Saxena A, Tang D, de Hoon M, Anelli V, Mione M, Carnici P, D'Adda di Fagagna F (2012) Site-specific DICER and DROSHA RNA products control the DNA-damage response. Nature. doi:10.1037/nature11179

14. Vagin VV, Sigova A, Li C, Seitz H, Gvozdev V, Zamore PD (2006) A distinct small RNA pathway silences selfish genetic elements in the germline. Science 313:320–324
15. Horwich MD, Li C, Matranga C, Vagin V, Farley G, Wang P, Zamore PD (2007) The Drosophila RNA methyltransferase, DmHen1, modifies germline piRNAs and single-stranded siRNAs in RISC. Curr Biol 17:1265–1272
16. Li C, Vagin VV, Lee S, Xu J, Ma S, Xi H, Seitz H, Horwich MD, Syrzycka M, Honda BM, Kittler EL, Zapp ML, Klattenhoff C, Schulz N, Theurkauf WE, Weng Z, Zamore PD (2009) Collapse of germline piRNAs in the absence of Argonaute3 reveals somatic piRNAs in flies. Cell 137:509–521
17. Seitz H, Ghildiyal M, Zamore PD (2008) Argonaute loading improves the 5′ precision of both MicroRNAs and their miRNA strands in flies. Curr Biol 18:147–151
18. Malone C, Brennecke J, Czech B, Aravin A, Hannon GJ (2012) Preparation of small RNA libraries for high-throughput sequencing. Cold Spring Harb Protoc 2012:1067–1077
19. Bentley DR, Balasubramanian S, Swerdlow HP, Smith GP, Milton J, Brown CG, Hall KP, Evers DJ, Barnes CL, Bignell HR, Boutell JM, Bryant J, Carter RJ, Keira Cheetham R, Cox AJ, Ellis DJ, Flatbush MR, Gormley NA, Humphray SJ, Irving LJ, Karbelashvili MS, Kirk SM, Li H, Liu X, Maisinger KS, Murray LJ, Obradovic B, Ost T, Parkinson ML, Pratt MR, Rasolonjatovo IM, Reed MT, Rigatti R, Rodighiero C, Ross MT, Sabot A, Sankar SV, Scally A, Schroth GP, Smith ME, Smith VP, Spiridou A, Torrance PE, Tzonev SS, Vermaas EH, Walter K, Wu X, Zhang L, Alam MD, Anastasi C, Aniebo IC, Bailey DM, Bancarz IR, Banerjee S, Barbour SG, Baybayan PA, Benoit VA, Benson KF, Bevis C, Black PJ, Boodhun A, Brennan JS, Bridgham JA, Brown RC, Brown AA, Buermann DH, Bundu AA, Burrows JC, Carter NP, Castillo N, Chiara ECM, Chang S, Neil Cooley R, Crake NR, Dada OO, Diakoumakos KD, Dominguez Fernandez B, Earnshaw DJ, Egbujor UC, Elmore DW, Etchin SS, Ewan MR, Fedurco M, Fraser LJ, Fuentes-Fajardo KV, Scott-Furey W, George D, Gietzen KJ, Goddard CP, Golda GS, Granieri PA, Green DE, Gustafson DL, Hansen NF, Harnish K, Haudenschild CD, Heyer NI, Hims MM, Ho JT, Horgan AM, Hoschler K, Hurwitz S, Ivanov DV, Johnson MQ, James T, Huw Jones TA, Kang GD, Kerelska TH, Kersey AD, Khrebtukova I, Kindwall AP, Kingsbury Z, Kokko-Gonzales PI, Kumar A, Laurent MA, Lawley CT, Lee SE, Lee X, Liao AK, Loch JA, Lok M, Luo S, Mammen RM, Martin JW, McCauley PG, McNitt P, Mehta P, Moon KW, Mullens JW, Newington T, Ning Z, Ling Ng B, Novo SM, O'Neill MJ, Osborne MA, Osnowski A, Ostadan O, Paraschos LL, Pickering L, Pike AC, Chris Pinkard D, Pliskin DP, Podhasky J, Quijano VJ, Raczy C, Rae VH, Rawlings SR, Chiva Rodriguez A, Roe PM, Rogers J, Rogert Bacigalupo MC, Romanov N, Romieu A, Roth RK, Rourke NJ, Ruediger ST, Rusman E, Sanches Kuiper RM, Schenker MR, Seoane JM, Shaw RJ, Shiver MK, Short SW, Sizto NL, Sluis JP, Smith MA, Ernest Sohna Sohna J, Spence EJ, Stevens K, Sutton N, Szajkowski L, Tregidgo CL, Turcatti G, Vandevondele S, Verhovsky Y, Virk SM, Wakelin S, Walcott GC, Wang J, Worsley GJ, Yan J, Yau L, Zuerlein M, Mullikin JC, Hurles ME, McCooke NJ, West JS, Oaks FL, Lundberg PL, Klenerman D, Durbin R, Smith AJ (2008) Accurate whole human genome sequencing using reversible terminator chemistry. Nature 456:53–59
20. Langmead B, Trapnell C, Pop M, Salzberg SL (2009) Ultrafast and memory-efficient alignment of short DNA sequences to the human genome. Genome Biol 10:R25

Chapter 5

Methods for Studying the Biological Consequences of Endo-siRNA Deficiency in *Drosophila melanogaster*

Do-Hwan Lim, Chun-Taek Oh, Sung-Jun Han, and Young Sik Lee

Abstract

Endogenous small interfering RNAs (endo-siRNAs) are a newly emerged class of small regulatory RNAs. In *Drosophila melanogaster*, the production of endo-siRNAs depends on the RNase III enzyme Dicer-2 (Dcr-2). Loss of *dcr-2* function reduces the resistance of adult flies to various stresses and shortens their life span. The mutants also exhibit alterations in carbohydrate and lipid metabolism. These findings suggest that the endo-siRNA pathway plays a protective role during exposure to stress and aging in *D. melanogaster*, possibly by regulating metabolic homeostasis. Here, we describe the methods that were used to discover the phenotypes associated with endo-siRNA deficiency by using *dcr-2* null mutants.

Key words *Drosophila melanogaster*, Endogenous siRNAs, Dicer-2, Stress resistance, Life span, Metabolism

1 Introduction

Small noncoding RNAs mostly function to repress cognate transcripts in multicellular organisms. The small regulatory RNAs can be categorized into three fundamental classes: 21-nucleotide (nt) endogenous small interfering RNAs (endo-siRNAs), ~22-nt microRNAs (miRNAs), and ~24–32-nt Piwi-interacting RNAs (piRNAs). These small RNAs are involved in a wide range of biological processes, the maintenance of genome integrity, and antiviral defense [1]. In *Drosophila melanogaster*, the three classes of small RNAs are produced through distinct biogenesis pathways and associate with specialized Argonaute (AGO)-family proteins. The production of both endo-siRNAs and miRNAs is dependent on the RNase III enzyme Dicer. Although only a single Dicer is found in *Caenorhabditis elegans* and mammals, *D. melanogaster* has two Dicer paralogs, Dicer-1 (Dcr-1) and Dicer-2 (Dcr-2), with specialized functions—Dcr-1 produces miRNAs from short hairpin RNA precursors, whereas Dcr-2 generates endo-siRNAs from double-stranded RNA precursors, e.g., transcripts with extended hairpin

Andreas Werner (ed.), *Animal Endo-siRNAs: Methods and Protocols*, Methods in Molecular Biology, vol. 1173, DOI 10.1007/978-1-4939-0931-5_5,

structures and sense–antisense transcript hybrids [1, 2]. Similarly, mature miRNAs and endo-siRNAs are loaded into functionally distinct AGO-subfamily proteins (AGO1 and AGO2, respectively) to form effector complexes targeting transcripts complementary to the small RNA [2]. Unlike miRNAs and endo-siRNAs, piRNAs are generated through a Dicer-independent pathway and are bound by AGO-proteins of the Piwi-subfamily to repress transposable elements [1, 2].

Endo-siRNAs have been identified in *C. elegans*, *D. melanogaster*, and mammals, with the advent of deep sequencing technology [1, 2]. These endo-siRNAs have extensive complementarity to mRNAs and transposable elements, which indicates their roles in repressing both protein-coding genes and transposons. Consistent with this idea, a deficiency of endo-siRNAs leads to an increase in cognate transcript levels [2]. However, the specific biological functions of endo-siRNAs in animals remain poorly understood.

D. melanogaster offers an advantage over other animal species as a model organism for studying the biological and physiological relevance of the endo-siRNA pathway because it encodes functionally distinct Dicer enzymes [2, 3]. Our recent work on phenotypic analysis of *dcr-2* null mutants with a significant defect in endo-siRNA production suggested that the endo-siRNA pathway is involved in metabolic regulation and has an impact on stress resistance and life span in *D. melanogaster* [4]. In the following sections, we describe several of the methods used for the phenotypic analysis of *dcr-2* mutants. These methods are also applicable to study gain- or loss-of-function phenotypes of other genes in *D. melanogaster*.

2 Materials

Prepare all solutions using ultrapure water (18 MΩ dH_2O) and analytical grade reagents. Maintain flies in an incubator under constant conditions—temperature, 25 °C; 65 % humidity; and 12-h light–dark cycle—unless otherwise stated. Sort adult flies by sex under a dissecting microscope by using brief CO_2 anesthesia. Dispose waste materials properly by following local regulations. The materials and reagents used in the various tests and procedures are listed below.

2.1 Drosophila Culture

1. *D. melanogaster* mutant lines homozygous for two independent *dcr-2* null alleles (*y w eyFLP*; *FRT42D*, *dcr-2*^{R416X} and *y w eyFLP*; *FRT42D*, *dcr-2*^{L811fsX}), their parental strain (*y w eyFLP*; *FRT42D*), and the *dcr-2*^{L811fsX} line bearing two copies of an amino-terminal hemagglutinin (HA)-tagged version of the *dcr-2* genomic rescue transgene, which is referred to as *HA-dcr-2* (*y w eyFLP*; *FRT42D*, *dcr-2*^{L811fsX}; *HA-dcr-2*) (*see* **Note 1**).

2. *Drosophila* medium: 6.97 % cornmeal, 2 % yeast, 9.6 % sugar, 1 % agar, 0.03 % methyl 4-hydroxybenzoate, and 0.4 % propionic acid (*see* **Note 2**). Hydrate 120 g baker's yeast with 3 L distilled water. Add 418.2 g cornmeal to the yeast slurry and mix well. Weigh 576 g black sugar and 60 g agar powder and hydrate with 1.5 L distilled water. Mix the yeast–cornmeal and sugar–agar mixtures together, and make the volume up to 6 L with distilled water. Transfer the combined mixture to a large electric kettle and boil with continuous stirring until the agar is dissolved (*see* **Note 3**). Cool the mixture to 60 °C and add 24 mL propionic acid and 18 mL 10 % methyl 4-hydroxybenzoate in 95 % ethanol. Dispense 3 and 30 mL of the *Drosophila* medium into fly culture vials (25 × 95 mm) and bottles (57 × 57 × 103 mm), respectively. Cover the vials and bottles with cotton gauzes and cool them to room temperature before closing with cotton plugs. Store at 4 °C in plastic bags (*see* **Note 4**).

2.2 Oxidative and Endoplasmic Reticulum (ER) Stress Tests

1. Oxidative stress solution: 5 % sucrose, 10 mM Paraquat. Dissolve 0.5 g sucrose and 25.7 mg Paraquat (methyl viologen dichloride hydrate, Sigma-Aldrich) in water up to 10 mL. Prepare fresh each time.
2. ER stress solution: 5 % sucrose, 100 mM dithiothreitol (DTT). Dissolve 0.5 g sucrose and 154.3 mg DTT (Sigma-Aldrich) in water up to 10 mL. Prepare fresh each time.
3. Sucrose solution (5 %). Dissolve 0.5 g sucrose in water up to 10 mL. Prepare fresh each time.

2.3 Measurement of Total Hemolymph Glucose in Larvae

1. Vials containing 2 % agar. Add approximately 400 mL distilled water to 10 g agar powder. Heat the mixture in a microwave with occasional swirling until the agar is completely dissolved. Make the volume up to 500 mL with hot distilled water and cool the mixture to 60 °C. Dispense 3 mL of the agar into fly culture vials and store as described in Subheading 2.1 (*see* **Note 4**).
2. Forceps with "Biologie" tips (Dumont #5).
3. Glucose Hexokinase Reagent supplemented with trehalase. Reconstitute Glucose Hexokinase Reagent (Thermo Scientific) in 50 mL water. Store at 4 °C. Add 10 μL porcine kidney trehalase (Sigma-Aldrich) to 5 mL Glucose Hexokinase Reagent (*see* **Note 5**). Mix and adjust the pH to 6.8 by using 1 N HCl. Prepare fresh each time.
4. Glucose (Sigma-Aldrich).

2.4 Triglyceride Measurement

1. Lysis solution: 10 mM Tris–HCl, pH 8.0, 150 mM NaCl, 0.1 % Triton X-100. Add approximately 90 mL water to 0.121 g Tris. Mix and adjust to pH 8.0 with HCl. Add 0.877 g NaCl

and 100 μL Triton X-100. Mix and make the volume up to 100 mL with water. Autoclave the solution and store at 4 °C.

2. Tissue-Tearor (BioSpec Products).
3. Triglycerides Reagent (Thermo Scientific).

3 Methods

3.1 Oxidative and ER Stress Tests

1. Prepare 9–10-day-old adult flies of each genotype (*see* **Note 6**).
2. Place groups of 60 flies into empty vials by using brief CO_2 anesthesia and starve them for 3 h at 25 °C.
3. During the dry starvation period, prepare vials containing a Whatman 3 MM filter paper disk soaked with 150 μL of 5 % sucrose–10 mM Paraquat (oxidative stress) solution, 5 % sucrose–100 mM DTT (ER stress) solution, or 5 % sucrose alone (*see* **Note 7**).
4. Transfer male and female flies of each genotype in groups of 20 to vials with or without Paraquat or DTT (*see* **Note 8**).
5. Culture the flies in the dark at 25 °C and count the dead flies in each group every hour until all the flies are dead (*see* **Note 9**).

3.2 Dry Starvation Test

1. Prepare 9–10-day-old adult flies of each genotype (*see* **Note 6**).
2. Place groups of 20 flies of each sex into empty vials.
3. Count the dead flies in each group at 25 °C every hour until all the flies are dead.

3.3 Cold Stress Test

1. Prepare 9–10-day-old adult flies of each genotype (*see* **Note 6**).
2. Transfer flies of each sex in groups of 20 to vials containing the *Drosophila* medium.
3. Culture the flies in a daily cycle of 8 h at 4 °C with a 16-h recovery period at 25 °C.
4. Count the dead flies every day until all the flies are dead, while transferring survivors of each group every 2–3 days to vials containing fresh *Drosophila* medium without anesthesia.

3.4 Life Span Analysis

1. Prepare adult flies of each genotype within 24 h post-eclosion (*see* **Note 6**).
2. Allow 1 day for the flies to mate at 25 °C.
3. Transfer groups of 30 flies of each sex to vials containing the *Drosophila* medium.
4. Culture the flies at 25 °C and transfer flies of each group to vials containing fresh *Drosophila* medium without anesthesia every 2–3 days.
5. Record the number of dead flies in each group every 24 h until all the flies are dead.

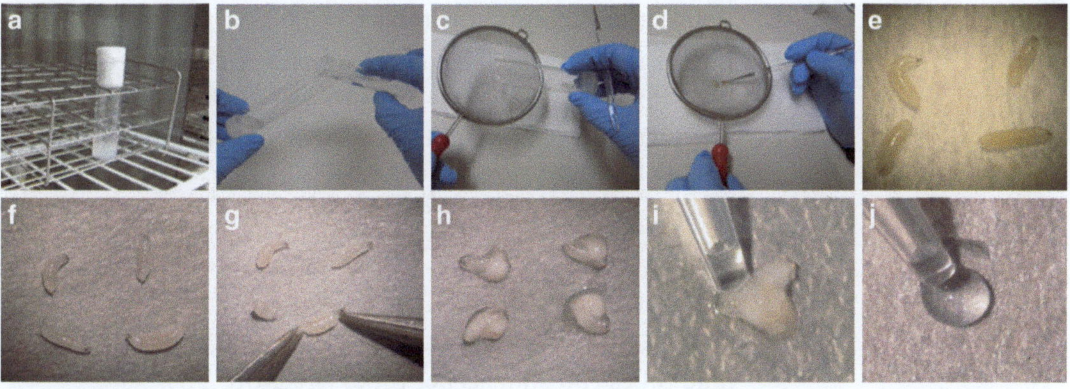

Fig. 1 Collection of larval hemolymph. (**a**) Starve synchronized third-instar larvae of each genotype on 2 % agar. (**b**) Release the larvae from the agar by wetting it slightly with distilled water and brushing softly. (**c**) Harvest the larvae by using a stainless steel mesh. (**d**) Wash the larvae with distilled water and blot dry using paper towels. (**e**) Transfer the larvae of each genotype in groups of four to Kimwipes papers to remove any residual water. (**f**) Transfer the larvae of each group to pieces of Parafilm. (**g, h**) Softly tear the cuticle away by pulling both ends of each larva lengthwise by using two pairs of forceps to bleed the hemolymph out. (**i**) Collect the hemolymph from each larva. (**j**) Pool the hemolymph for each group

3.5 Measurement of Total Hemolymph Glucose in Larvae

1. Prepare synchronized third-instar larvae of each genotype (*see* **Note 6**).
2. Transfer approximately 30 larvae to a vial containing 2 % agar by using a soft brush.
3. Starve the larvae for 5 h at 25 °C (*see* **Note 10**) (Fig. 1a).
4. Release the larvae from the agar by wetting it slightly with distilled water and brushing softly (Fig. 1b).
5. Harvest the larvae of each genotype by using a stainless steel mesh, wash thoroughly with distilled water, and remove excess water (Fig. 1c, d).
6. Transfer the larvae in groups of four to Kimwipes papers to remove any residual water (Fig. 1e).
7. Transfer the larvae of each group to pieces of Parafilm (Fig. 1f).
8. Grab the anterior and posterior ends of the larva by using two pairs of sharp forceps, and pull the larva gently lengthwise to tear the cuticle away, thereby allowing the hemolymph to bleed out (Fig. 1g, h).
9. Collect the hemolymph from each larva and pool for each group (*see* **Note 11**) (Fig. 1i, j).
10. Mix 0.4 μL of the pooled hemolymph with 200 μL Glucose Hexokinase Reagent supplemented with trehalase in a 1.5-mL microcentrifuge tube.
11. Incubate the mixture at 37 °C overnight (*see* **Notes 5** and **12**).

12. Transfer the reaction mixtures to a 96-well plate.
13. Measure the absorbance of the samples at 340 nm relative to a blank by using a microplate reader (*see* **Note 13**).
14. Calculate the amount of total hemolymph glucose in each sample by comparing the absorbance with a standard curve for glucose concentration (*see* **Note 14**).

3.6 Whole-Body Triglyceride Measurement

1. Prepare synchronized third-instar larvae or 9–10-day-old male and female flies of each genotype (*see* **Note 6**).
2. Transfer the larvae or flies of each sex to 1.5-mL microcentrifuge tubes (30 animals per tube) (*see* **Note 15**).
3. Homogenize animals of each group in 600 μL lysis solution by using a Tissue-Tearor on ice.
4. Filtrate the homogenate by using a 0.45-μm filter to remove debris.
5. Mix 7 μL of each sample with 150 μL Triglycerides Reagent in a 96-well plate.
6. Incubate the reaction mixtures at 37 °C for 10 min (*see* **Note 16**).
7. Measure the absorbance of the samples at 500 nm relative to a blank by using a microplate reader (*see* **Note 13**).
8. Normalize the triglyceride levels to the amount of protein in each sample (*see* **Note 17**).

4 Notes

1. The *dcr-2* null mutants and their parental strain were described originally by Lee et al. [3]. The *dcr-2*^{L811fsX} line bearing two copies of the *HA-dcr-2* transgene was described by Lim et al. [4].
2. All dry ingredients should be completely hydrated with water by using a hand mixer to mix until lump-free.
3. Once the agar is completely dissolved, turn off the heat to avoid excess evaporation of water.
4. Before use, the vials or bottles should be pre-warmed to room temperature to protect the larvae or adult flies from cold stress.
5. Hemolymph contains trehalose (a disaccharide of glucose) and, to a lesser extent, monomeric glucose as circulating sugars, which are referred to as total hemolymph glucose. One molecule of trehalose is hydrolyzed to two molecules of glucose in the reaction catalyzed by trehalase.
6. To generate sufficient numbers of experimental flies, place healthy parents of each genotype into several bottles containing the *Drosophila* medium supplemented with yeast paste. The parent flies lay eggs within ~2 h at 25 °C. Thereafter, estimate

the number of the eggs to ensure synchronized development and growth of the animals at a rearing density of ~300 flies per bottle. After removing the parent flies, maintain the bottles at 25 °C to permit further development of the eggs. For life span analysis, collect adult flies of each genotype and both sexes into bottles containing the *Drosophila* medium within 24 h of eclosion by using brief CO_2 anesthesia. For experiments with 9–10-day-old adult flies, collect flies by using the method described for the life span analysis, and age the flies for 9 days at 25 °C with transferring them to bottles containing fresh *Drosophila* medium without anesthesia every 2–3 days. For experiments with third-instar larvae, wet the *Drosophila* medium in a rearing bottle containing larvae of each genotype with ~3 mL of distilled water at 4 days after egg laying at 25 °C and brush softly to release the larvae from the medium. Harvest the larvae by using a stainless steel mesh (fine enough to collect them). Wash the larvae with tap water and then with distilled water and blot them dry using paper towels.

7. Prior to oxidative and ER stress tests, flies should be starved to avoid the variation arising from differential ingestion rates of paraquat or DTT. A sufficient volume of the solution should be absorbed by the filter paper disk such that drying during the test is minimized without interrupting the locomotor activity of the flies. A solution containing 5 % sucrose alone is used to monitor whether dry starvation itself compromises the viability of flies of each genotype under the same experimental conditions as those used for the oxidative and ER stress tests.
8. Anesthetized flies should be transferred carefully to the vials to prevent them from being stuck to the wet filter paper disk while unconscious.
9. Paraquat and DTT are light sensitive.
10. Starvation of third-instar larvae before collecting the hemolymph is recommended in order to synchronize the digestion status of the larvae, thereby reducing the effect of outliers.
11. The pooled hemolymph should be clear and transparent. Otherwise, it may be contaminated with larval tissue or the contents of organs and should be discarded.
12. A two-step reaction catalyzes the conversion of glucose to 6-phosphogluconate, producing a reduced form of nicotinamide adenine dinucleotide (NADH). The resulting NADH is detected by measuring the absorbance at 340 nm; therefore, an increase in the absorbance is proportional to the concentration of glucose present in the sample.
13. The blank should be prepared using the same solution as in the samples.

14. A standard curve should be generated for each trial to determine the amount of glucose in a linear range of 0–100 mg/mL. Prepare glucose standard solutions (100, 50, 25, 12.5, and 6.25 mg/mL) in phosphate-buffered saline by using twofold serial dilutions starting at 100 mg/mL. The end volume of all solutions is 200 μL. Proceed to **steps 10–13** of Subheading 3.5 by using 0.4 μL of each standard solution instead. Plot the standard curve with the glucose concentration on the *X*-axis and the absorbance on the Y-axis. Calculate the glucose concentration in unknown samples by using linear regression.
15. Use the larvae or flies as fed animals. To measure triglyceride content under starvation conditions, the animals should be maintained on 2 % agar for a given period at 25 °C, prior to homogenization.
16. In a series of reactions, triglycerides are converted to dihydroxyacetone phosphate, thereby generating hydrogen peroxide. A subsequent color reaction produces a red-colored dye from the resulting hydrogen peroxide. The absorbance of the dye at 500 nm is proportional to the concentration of triglycerides in the sample.
17. The amount of protein in each sample can be quantified using the Bradford method [5].

Acknowledgements

This work was supported by a grant from the National Research Foundation of Korea (NRF) funded by the Korea government (Ministry of Education, Science, and Technology [MEST]; 2011-0016863 to Y.S.L.) and the NRF grant funded by MEST (2012-00011, Gyeonggi-do and KISTI to S.J.H.).

References

1. Kim VN, Han J, Siomi MC (2009) Biogenesis of small RNAs in animals. Nat Rev Mol Cell Biol 10:126–139
2. Okamura K, Lai EC (2008) Endogenous small interfering RNAs in animals. Nat Rev Mol Cell Biol 9:673–678
3. Lee YS, Nakahara K, Pham JW, Kim K, He Z, Sontheimer EJ, Carthew RW (2004) Distinct roles for Drosophila Dicer-1 and Dicer-2 in the siRNA/miRNA silencing pathways. Cell 117:69–81
4. Lim DH, Oh CT, Lee L, Hong JS, Noh SH, Hwang S, Kim S, Han SJ, Lee YS (2011) The endogenous siRNA pathway in Drosophila impacts stress resistance and lifespan by regulating metabolic homeostasis. FEBS Lett 585: 3079–3085
5. Bradford MM (1976) A rapid and sensitive method for the quantitation of microgram quantities of protein utilizing the principle of protein-dye binding. Anal Biochem 72:248–254

Chapter 6

Small RNA Library Cloning Procedure for Deep Sequencing of Specific Endogenous siRNA Classes in *Caenorhabditis elegans*

Maria C. Ow, Nelson C. Lau, and Sarah E. Hall

Abstract

In recent years, distinct classes of small RNAs ranging in size from ~21 to 26 nucleotides have been discovered and shown to play important roles in a wide array of cellular functions. Because of the abundance of these small RNAs, library preparation from an RNA sample followed by deep sequencing provides the identity and quantity of a particular class of small RNAs. In this chapter we describe a detailed protocol for preparing small RNA libraries for deep sequencing on the Illumina platform from the nematode *C. elegans*.

Key words Endogenous siRNAs, *C. elegans*, RNAi, 26G-siRNA, 22G-siRNA, Small RNA library cloning

1 Introduction

Since the description of the first regulatory noncoding RNA cloned from the nematode *C. elegans* two decades ago, the field of small RNAs (sRNAs) has been implicated in a myriad of developmental and gene regulation processes in eukaryotes spanning from yeast to humans [1]. In animals, three major types of small RNAs have been identified: microRNAs (miRNAs), PIWI-interacting RNAs (piRNAs), and endogenous short interfering RNAs (endo-siRNAs) (Fig. 1). MicroRNAs are generated by processing single-stranded RNA hairpin loops into ~21–22 nucleotides sRNAs through the action of two RNase III-like enzymes, Dicer and Drosha, and RNA-binding protein Pasha/DGCR8. MicroRNAs suppress translation and promote degradation of their target mRNAs by imperfectly base-pairing to their 3′ untranslated regions [2]. Both miRNAs and piRNAs are generated from single-stranded RNA precursors transcribed from genomic regions, although their biogenesis and respective function differ significantly. While the 5′ ends of both miRNAs and piRNAs have a 5′ monophosphate, their

Andreas Werner (ed.), *Animal Endo-siRNAs: Methods and Protocols*, Methods in Molecular Biology, vol. 1173,
DOI 10.1007/978-1-4939-0931-5_6, © Springer Science+Business Media New York 2014

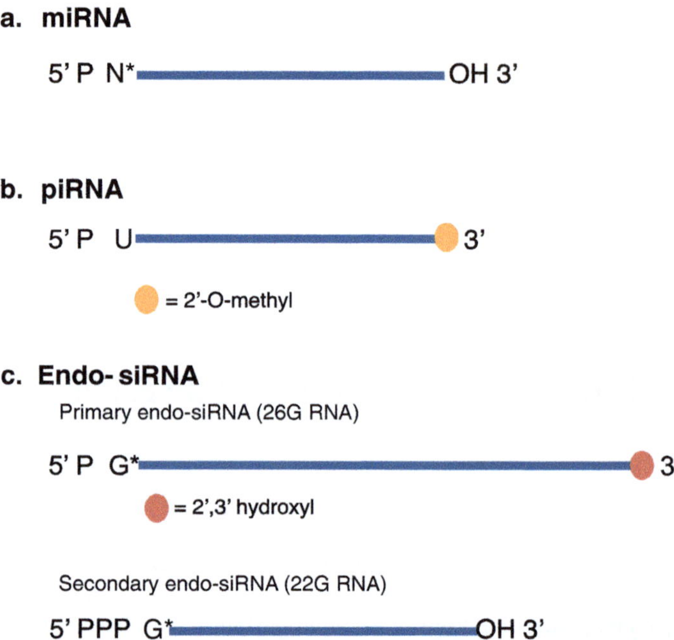

Fig. 1 The three major classes of sRNAs in *C. elegans*. (**a**) miRNAs (~21–22 nucleotides long) are Dicer-dependent and thus have a 5′ monophosphate and a 3′ hydroxyl modifications. Most (>70 %) miRNAs start with a uridine at the 5′ end (indicated as N*). The ~30 % of the remaining miRNAs start with an adenosine, cytidine, or guanosine. (**b**) 21U-piRNAs exhibit a 5′ monophosphate uracil and a 2′-methyl moiety at the 3′ end. (**c**) Endo-siRNAs are categorized as primary endo-siRNAs (26G-siRNAs) and secondary endo-siRNAs (22G-siRNAs). The lesser abundant 26G-siRNAs are Dicer-dependent and have a 5′ monophosphate guanosine and a 2′,3′ hydroxyl group at the 3′ end. The more abundant 22G-siRNAs are Dicer-independent, RdRP-dependent, and have a 5′ triphosphate guanosine and a 3′ hydroxyl group. The use of TAP (*see* Subheading 3.1) allows for the conversion of 5′ triphosphate to a monophosphate group required for sRNA library cloning following this protocol. G* indicates that over 90 % of endo-siRNAs have a guanosine at the 5′ end. The remaining ~10 % of endo-siRNAs can start with an adenosine, cytidine, or uridine

3′ ends differ significantly with miRNAs having a 3′ OH and piRNAs possessing a 2′-O-methyl moiety (Fig. 1). Unlike miRNAs, production of piRNAs is Dicer-independent and involves a "ping pong" amplification mechanism. All *C. elegans* piRNAs bind to the PIWI-containing clade of Argonautes (AGOs) and possess a 5′ uracil (Fig. 1); however, the length of piRNAs varies from 21 nt in *C. elegans* (also referred to as 21U-RNAs) to ~24–30 nt in *Drosophila*, zebrafish, and mammals. Although the original function of piRNAs was characterized as transposon silencing in the germ line, recent identified functions in *C. elegans* include regulation of chromatin structure and transgenerational cellular memory of "self" versus "nonself" [3–5].

Of the three major classes of sRNAs in *C. elegans* adults, endo-siRNAs are the most abundant and variable class. Endo-siRNAs range from ~21 to 26 nucleotides in length and are most often antisense to coding regions of transposons and endogenous genes, although endo-siRNAs mapping to intergenic regions such as the X-cluster have been identified. One method of characterizing this class of sRNAs is by their biogenesis. Primary endo-siRNAs are generated from double-stranded RNA processed by Dicer, which results in a siRNA containing a characteristic 5′ monophosphate and 2′,3′ hydroxy terminus (Fig. 1). Primary endo-siRNAs are of low abundance, ~26 nucleotides in length, and possess a 5′ guanosine (26G-siRNAs) (Fig. 1). Alternatively, secondary endo-siRNAs are highly abundant, are ~22 nucleotides in length, and possess a 5′ guanosine (22G-siRNAs). In contrast to 26G-siRNAs, the biogenesis of 22G-siRNAs is through the action of RNA-dependent RNA polymerases (RdRPs) and is Dicer-independent, resulting in a 5′ triphosphate cap (Fig. 1) [6, 7]. While the biogenesis of secondary siRNAs in worms and plants is dependent on RdRPs, no RdRPs have been identified in flies and mammals, suggestive of an RdRP-independent biogenesis mechanism [2]. Although evidence suggests that secondary endo-siRNA biogenesis in *C. elegans* is stimulated by the primary 26G-endo-siRNAs, the mechanisms regulating this process remain unclear [8].

The use of exogenous dsRNA to "knockdown" expression of a target gene locus through RNAi has become routine practice in many model organisms; however, the function of endogenous siRNAs is less understood. The *C. elegans* genome contains 26 AGO genes that play a role in the regulation of gene expression and chromatin state. The large number of *C. elegans* AGOs, a majority of which bind endo-siRNAs, allows for functional specificity within worms based on endo-siRNA biogenesis, tissue specificity, and site of action [9, 10]. For example, NRDE-3 AGO is expressed in somatic tissue, binds to 22G-siRNAs, and targets genomic loci in the nucleus to regulate chromatin state. Alternatively, ALG-3/4 AGOs are expressed in sperm, bind 26G-endo-siRNAs and target sperm-enriched mRNAs to regulate thermotolerant male fertility [11–14]. In recent years with the prevalence of deep-sequencing technologies, much progress has been made identifying potential targets of individual AGO proteins by sequencing AGO-bound siRNAs or by identifying genes with a reduction in siRNA abundance antisense to their coding sequences in mutant backgrounds [11, 12, 14–17]. However, efforts to correlate endo-siRNA abundance with changes in target gene expression have only been moderately successful, suggesting that endogenous RNAi pathways may have more complex gene regulatory functions than the exogenous RNAi pathways [11, 12, 16–18].

In this chapter, we present a general protocol for preparing sRNA libraries from *C. elegans* for deep sequencing on the Illumina

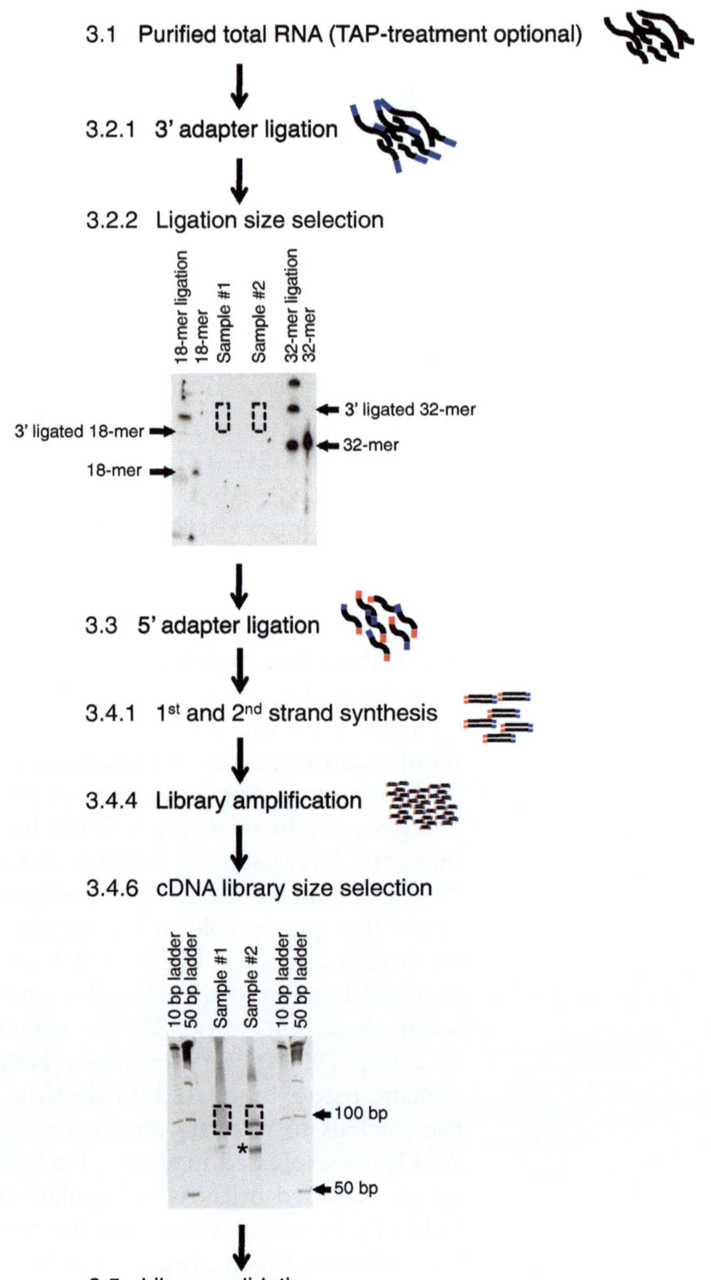

Fig. 2 Flowchart of sRNA library synthesis. This flowchart illustrates the steps of sRNA library synthesis outlined in this chapter. The numbers indicate which step in the protocol is being illustrated. Illustrative examples of polyacrylamide gels used for sRNA library size selection after completion of (**a**) Subheading 3.2 3′ adapter ligation and (**b**) Subheading 3.4 reverse transcription and library amplification steps are shown. The *dotted lines* indicate the regions of the gels that were extracted. *Asterisk* indicates the primer dimer artifact described in the text

platform (Fig. 2) (modified from ref. 19). This process requires purification of small RNAs from total RNA and ligation of adapter oligos to the 5′ and 3′ ends of the sRNAs. The ligation reactions are sensitive to the 5′ phosphate status of the sRNAs, allowing for enrichment of either primary endo-siRNAs (5′ monophosphate) or secondary endo-siRNAs (5′ triphosphate), if desired (Fig. 1). An optional enzymatic step using Tobacco Acid Phosphatase (TAP) hydrolyzes the phosphoric acid anhydride bonds in the 5′ triphosphorylated ends of 22G-endo-siRNAs (*see* Subheading 3.1), leaving them with a 5′ monophosphate that can be substrates for the ligation of the 5′ adapter oligo. Finally, we describe reverse transcription, amplification, and validation steps to prepare the sRNA libraries for deep sequencing. The protocol described here differs from the endo-siRNA sequencing protocols used for other metazoans. For instance, in *Drosophila melanogaster*, endo-siRNAs bear 5′-monophosphorylated and 2′-O-methylated 3′ ends. This feature is used to select against other sRNAs (e.g., miRNAs which are not modified at the 3′ end) by treating a total sRNA sample with $NaIO_4$ followed by β-elimination, resulting in the blockage of adapter ligation to sRNAs with a 2′,3′ hydroxyl termini and thus excluding them from being cloned [20]. In the future, endo-siRNA sequencing and analysis from specific tissue types, developmental stages, and environmental conditions will likely be required to fully elucidate the complex regulatory mechanisms of endogenous RNAi pathways.

2 Materials

Take proper precautions to avoid RNase contamination throughout the protocol, including using baked glassware and DEPC or Milli-Q-treated water to make RNase-free solutions.

2.1 Preparation of RNA

Trizol reagent (Invitrogen).

Tobacco acid pyrophosphatase (TAP) (Epicentre) (optional).

3 M Sodium acetate (NaOAc).

100 % Ethanol.

2.2 3′ Adapter Ligations and Purification of Samples

Denaturing and native polyacrylamide gel electrophoresis supplies (such as SequaGel).

5× Ligation buffer: 250 mM 2-[4[(2-hydroxyethyl)piperazin-1-yl] ethanesulfonic acid (HEPES) pH 8.3, 50 mM magnesium chloride ($MgCl_2$), 16.5 mM dithiothreitol (DTT), 50 μg/mL bovine serum albumin (BSA), and 41.5 % glycerol. Store at −20 °C.

T4 single-stranded RNA ligase enzyme.

RNase Inhibitor (optional).

PhosphorImager and exposure screens.

0.3 M Sodium chloride (NaCl).

Glycogen.

3′ adapter (for Hi-Seq 2000).

pCGTCGTATGCCGTCTTCTGCTTGT/3AmMO/.

This is adapter is phosphorylated at the 5′ end and contains a /3AmMO/ amino modifier at the 3′ end.

2.3 5′ Adapter Ligation of Experimental Samples

5′ adapters with barcode (for Hi-Seq 2000).

Barcode CAA: gttcagagttctacagtccgacgatcNNN**CAAAA**.

Barcode ACC: gttcagagttctacagtccgacgatcNNN**ACCAA**.

Barcode GUU: gttcagagttctacagtccgacgatcNNN**GUUAA**.

Barcode UGG: gttcagagttctacagtccgacgatcNNN**UGGAA**.

DNA bases are in lowercase, N is a random DNA base, and RNA bases are in uppercase bold with the unique barcode underlined.

5 mM Adenosine triphosphate (ATP).

2.4 Reverse Transcription and Library Amplification

5′ PCR primer (18.206): CAAGCAGAAGACGGCATA.

3′ PCR primer (44.45): AATGATACGGCGACCACCGACAG GTTCAGAGTTCTACAGTCCGA.

Reverse transcription kit with enzyme and buffer (such as SuperScript III, Invitrogen).

2 mM dNTPs.

High fidelity DNA polymerase enzyme with buffer (such as Phusion, New England Biolabs).

10 and 50 bp DNA ladders.

1× Tris–Borate–EDTA (TBE) buffer.

Ethidium bromide (EtBr).

10 mM Tris pH 8.

1× glycerol loading dye.

2.5 Library Validation (Optional)

Blunt-end or TA cloning kit.

2.6 Radioactive Labeling of RNA Oligo Size Markers

18mer RNA oligo: AGCGUGUAGGGAUCCAAA.

32mer RNA oligo: GGCAUUAACGCGGCCGCUCUACAAUA GUGA.

Polynucleotide kinase (PNK) enzyme with buffer.

^{32}P γ–Adenosine triphosphate (^{32}P γ–ATP).

2.7 3′ Adapter Synthesis

100 μM of 3′ adapter.

5′ DNA Adenylation Kit (New England Biolabs): 10× DNA Adenylation Buffer, 1 mM ATP, and Mth RNA Ligase.

3 Methods

Carry out all procedures at room temperature unless otherwise specified.

3.1 Preparation of RNA

1. Before beginning library preparation, isolate high-quality total RNA from *C. elegans* samples (*see* **Note 1**). If capture of all types of *C. elegans* endo-siRNAs is desired (*see* Subheading 1), proceed to treatment of the total RNA with tobacco acid pyrophosphatase (TAP) below. If cloning of Dicer-generated primary endo-siRNAs only is desired, proceed to Subheading 3.2.
2. TAP enzyme converts the 5′ triphosphate cap that is present on RdRP generated endo-siRNAs to a 5′ monophosphate that is compatible with the 5′ ligation procedure described here. Manufacturer's instructions should be followed for RNA treatment in a volume <100 μL (*see* **Note 2**).
3. Add water to the stopped-TAP treatment reaction for a total volume of 100 μL (*see* **Note 3**). Ethanol-precipitate treated RNA by first adding 10 μL 3 M sodium acetate and then 250 μL of 100 % ethanol, mixing well after addition of each reagent. Incubate the sample at −20 °C for at least 30 min, and then spin the sample in a microcentrifuge at the maximum speed for 10 min at 4 °C. You should be able to visualize a clear to white pellet. Remove supernatant and allow RNA pellet to dry for ~5 min. Resuspend RNA in 9 μL of water.

3.2 3′ Adapter Ligation and Purification of Samples

1. Assemble separate 3′ adapter ligation reactions with experimental samples and radiolabeled 18-mer and 32-mer RNA oligos (*see* Subheading 3.6 for generation of labeled size oligos and Subheading 3.7 for generation of 3′ adapters). For experimental samples, add 9 μL of total RNA (2–3 μg/μL), 1 μL of 100 μM 3′ adapter, 4 μL 5× ligation buffer (no ATP), 2 μL T4 single-stranded RNA Ligase, 1 μL RNase Inhibitor (optional), and H_2O up to 20 μL. For reactions with RNA marker oligos, modify components above to add the 18-mer or 32-mer oligos and adjust water accordingly. Incubate reaction at 18 °C for 2–4 h.
2. Gel-purify the adapter ligation reaction products on a 15 % polyacrylamide gel. Run lanes of unligated size markers in addition to 3′ adapter ligated size markers (*see* **Note 4** and Fig. 2). Run the gel at 30 W for ~1 h. Disassemble gel apparatus and adhere gel to plastic wrap. Expose the gel to a PhosphorImager

screen for ~15–30 min. Scan the PhosphorImager screen and print the image at 100 % zoom to act as a template for gel extraction. Place the plastic wrap and gel on top of the gel image. Using the gel image as a guide, cut out the gel region corresponding to the size range between 3′ adapter ligated 18-mer and 3′ adapter ligated 32-mer and put into separate 1.5 mL microcentrifuge tubes, using a new razor blade for each sample (*see* **Note 5** and Fig. 2). Also cut out the gel regions corresponding to the 3′ adapter ligated radiolabeled markers.

3. Elute RNA from gel slices overnight using 500 μL of 0.3 M NaCl per 1.5 mL microcentrifuge tube with shaking or rotation at 4 °C. Move the supernatant to a new microcentrifuge tube taking care to avoid the gel slices. Add 40 μg of glycogen and 1 mL of 100 % EtOH. Mix the reagents well and incubate at −20 °C for >1 h. Spin the samples in a microcentrifuge at maximum speed for 20 min at 4 °C. A whitish pellet should be visible. Carefully remove supernatant and air-dry pellet for ~5 min. Resuspend the RNA in 10 μL of H_2O. For the 18-mer and 32-mer ligations, resuspend the RNA pellets in 20 μL of H_2O.

3.3 5′ Adapter Ligation of Experimental Samples

Next-generation sequencing technology allows for the generation of several hundred million short DNA reads in a single run. The remarkable depth in the number of reads allows for the analysis of different experimental samples in parallel simply by the introduction of a unique identifier or barcode in the 5′ adapter that is ligated to the sRNAs during the cloning procedure.

1. To proceed with the 5′ adapter ligations, assemble the following reaction for each experimental sample and radiolabeled marker: 10 μL of purified 3′ ligation product, 4 μL 100 μM 5′ adapter, 4 μL 5× T4 RNA ligase buffer, 1 μL 5 mM ATP, and 1 μL T4 single-stranded RNA ligase. Incubate the reaction at 22 °C overnight.
2. Check for the completion of 5′ adapter ligation reaction by examining the ligation products of radiolabeled size markers on a 15 % polyacrylamide gel. Few side products are generated during this ligation reaction. Include the remaining 10 μL of 3′ adapter ligated products in separate lanes to verify the shift to a larger product in the 5′ and 3′ adapter ligated products. (*Optional*: You can also gel-purify the experimental sample ligations at this step if desired.) Run the gel at 30 W for ~1 h. If the 5′ adapter ligation reaction was successful, then ethanol-precipitate (or gel-extract) ligated experimental RNA products with glycogen. Resuspend pellet in 10 μL H_2O (*see* Subheading 3.2, **step 3**).

3.4 Reverse Transcription and Library Amplification

1. On ice, assemble a first strand reverse transcription reaction for each experimental sample. For best results, incubate reactions in a thermocycler. For each reaction, add in order: 10 μL purified 5′, 3′ adapter-ligated RNA sample, 1 μL 100 μM 5′ PCR primer, and 5 μL RNase-free H_2O. Heat the reaction to 65 °C for 5 min and cool on ice. Centrifuge the reactions to collect condensation before proceeding.
2. On ice, add the following reagents in order to the cooled reactions: 6 μL 5× first strand buffer, 6 μL 2 mM dNTPs, 2 μL 100 mM DTT, and 1 μL reverse transcriptase enzyme. Incubate the reaction at 50 °C for 1 h, 85 °C for 15 min, then cool to 4 °C (*see* **Note 6**).
3. After reverse transcription reaction is completed, immediately proceed to second strand extension. On ice, assemble a PCR reaction with the following reagents: 5 μL of first strand reverse transcription reaction from Subheading 3.4, **step 2**, 10 μL 5× PCR buffer, 6 μL 2 mM dNTPs, 2 μL 150 nM 3′ PCR primer, 25 μL RNase-free H_2O, and 1 μL high fidelity DNA polymerase. Perform second strand extension under the following conditions: 98 °C for 3 min, 5 cycles of 94 °C for 45 s, 60 °C for 45 s, and 72 °C for 45 s, then a final extension of 72 °C for 10 min. Cool the reaction to 10 °C in the thermocycler or on ice.
4. Next, add the following additional reagents to each reaction for library amplification: 10 μL 5× PCR buffer, 6 μL 2 mM dNTPs, 0.5 μL 25 μM 3′ PCR primer, 0.5 μL 25 μM 5′ PCR primer, 25 μL H_2O, and 1 μL high fidelity DNA polymerase. Perform library amplification under the following conditions: 98 °C for 3 min, 15–20 cycles of 94 °C for 30 s, 60 °C for 30 s, and 72 °C for 15 s, then a final extension of 72 °C for 10 min. Cool samples to 10 °C in the thermocycler or on ice.
5. Precipitate cDNA by adding 40 μg glycogen, 10 μL 3 M NaOAc, and 200 μL 100 % EtOH, mixing after addition of each reagent. Incubate the samples at −20 °C for >1 h, then spin at maximum speed in a microcentrifuge for 20 min at 4 °C. Remove supernatant and air-dry pellet for ~5 min. Resuspend cDNA in 15 μL of 1× glycerol loading dye.
6. Gel-extract the cDNA library using a 10 % native polyacrylamide gel. Load the samples near 10 and 50 bp DNA ladders, leaving empty lanes adjacent to the experimental samples to prevent contamination during gel extraction (Fig. 2). Run the gel in 1× TBE buffer at constant 25 W until the dye front reaches the bottom of the gel. Stain the gel with 4 μg/mL EtBr in 0.5× TBE buffer. Drain the EtBr stain from the gel and image on a gel documentation system. Using the gel image as a guide, cut out the gel pieces corresponding to 85–110 bp and place in separate microcentrifuge tubes (*see* **Note 7**).

7. Elute the cDNA library from the gel slices with 3 M NaCl overnight at room temperature (*see* Subheading 3.2, **step 3**). Ethanol-precipitate the cDNA with glycogen and resuspend in 15 μL of 10 mM Tris pH 8. Quantitate the concentrations of small RNA libraries. The desired concentration is >20 ng/μL with a total >200 ng DNA.

3.5 Library Validation (Optional)

1. Following the instructions of the blunt-end or TA ligation cloning kit, clone 1 μL of purified cDNA library sample for further validation. Purify the plasmid from 12 to 24 bacterial colonies per library samples and analyze by Sanger sequencing.
2. Use NCBI BLAST to determine sequence contents of plasmids. Libraries should contain clones of desired sRNAs to warrant sequencing.

3.6 Radioactive Labeling of RNA Oligo Size Markers

1. Assemble a separate polynucleotide kinase (PNK) reaction for each RNA oligo size marker: 1 μL of 10 μM RNA oligo (32mer or 18mer), 2 μL of 10× PNK buffer, 14 μL of RNase-free H_2O, 2 μL ^{32}P γ-ATP, and 1 μL T4 PNK (10 U/μL). Incubate reactions for 1 h at 37 °C.
2. Purify radiolabeled RNA oligos using a 15 % polyacrylamide gel. Run the gel at a constant 25 W in 0.5× TBE buffer until bromophenol blue dye front migrates to 1 in. from the bottom to eliminate excess radioactive ATP. Gel-extract and elute labeled RNA oligos (*see* Subheading 3.2, **step 3**). Resuspend the gel-purified RNA oligos in 20 μL H_2O.

3.7 3′ Adapter Synthesis

1. The 3′ adapter used for the preparation of sRNA libraries is a modified DNA oligonucleotide containing a 5′,5′-adenylpyrophosphoryl moiety and a blocking group (e.g., amine or dideoxynucleotides) at the 3′ end. The use of pre-adenylated oligodeoxynucleotides in the 3′ adapter ligation step allows the adapter to be ligated to the sRNAs by T4 RNA ligase in the absence of ATP, thus preventing the sRNA from self-ligating or concatamerizing. While these adenylated DNA oligos can be purchased from a vendor, synthesizing them as described below can be more cost-effective.
2. Adenylate the 3′ adapter following the instructions of the 5′ DNA Adenylation Kit: 100 μM of 3′ adapter, 2 μL of 10× DNA Adenylation Buffer, 2 μL of 1 mM ATP, 2 μL of Mth RNA Ligase (100 pmol), and RNase-free water to a volume of 20 μL.
3. Incubate the reaction at 65 °C for 1 h. Inactivate the enzyme by incubating at 85 °C for 5 min.

4 Notes

1. Trizol reagent (Invitrogen) works well to isolate high-quality RNA from *C. elegans* samples. The 260–280 absorbance ratio should be between 1.8 and 2.0 for high-quality RNA samples.
2. Starting with 20–30 μg of total RNA is ideal for small RNA gel extraction, and should be treated by TAP enzyme if the capture of all endo-siRNAs is desired.
3. Purification of the RNA by phenol–chloroform extraction is not necessary.
4. During this ligation step, side products are generated from ligations of just the radiolabeled oligo markers to themselves. Run unligated markers in separate lanes from the 3′ adapter ligated oligos to assist in size selection of RNA from the experimental samples. The side products will migrate faster than the unligated marker, or will appear too large (two markers concatamerizing).
5. Elution of RNA from the gel can be facilitated by cutting the larger slice into smaller pieces. However, make sure that the slices remain large enough so can be easily avoided while pipetting.
6. Instructions for the reverse transcription and PCR reactions described in this chapter are based on the optimized protocol for Superscript III reverse transcriptase (Invitrogen) and Phusion DNA polymerase (New England Biolabs). If using different enzymes, follow the manufacturer's instructions for the reactions.
7. If the starting material from the reverse transcription reaction is low, often a primer-dimer artifact (~70 bp long) will amplify and dominate the reaction, which can be identified during the gel purification step (Fig. 2). If this occurs, gel-purify the desired products between 85 and 110 bp, and use half of the resuspended sample for further PCR reamplification starting from Subheading 3.4, **step 4**.

References

1. Sabin LR, Delas MJ, Hannon GJ (2013) Dogma derailed: the many influences of RNA on the genome. Mol Cell 49:783–794
2. Kim VN, Han J, Siomi MC (2009) Biogenesis of small RNAs in animals. Nat Rev Mol Cell Biol 10:126–139
3. Ghildiyal M, Zamore PD (2009) Small silencing RNAs: an expanding universe. Nat Rev Genet 10:94–108
4. Shirayama M, Seth M, Lee HC et al (2012) piRNAs initiate an epigenetic memory of nonself RNA in the *C. elegans* germline. Cell 150:65–77
5. Ashe A, Sapetschnig A, Weick EM et al (2012) piRNAs can trigger a multigenerational epigenetic memory in the germline of *C. elegans*. Cell 150:88–99
6. Pak J, Fire A (2007) Distinct populations of primary and secondary effectors during RNAi in *C. elegans*. Science 315:241–244
7. Sijen T, Steiner FA, Thijssen KL et al (2007) Secondary siRNAs result from unprimed RNA

synthesis and form a distinct class. Science 315:244–247

8. Pak J, Maniar JM, Mello CC et al (2012) Protection from feed-forward amplification in an amplified RNAi mechanism. Cell 151: 885–899
9. Boisvert ME, Simard MJ (2008) RNAi pathway in *C. elegans*: the argonautes and collaborators. Curr Top Microbiol Immunol 320:21–36
10. van Wolfswinkel JC, Ketting RF (2010) The role of small non-coding RNAs in genome stability and chromatin organization. J Cell Sci 123:1825–1839
11. Gent JI, Lamm AT, Pavelec DM et al (2010) Distinct phases of siRNA synthesis in an endogenous RNAi pathway in *C. elegans* soma. Mol Cell 37:679–689
12. Conine CC, Batista PJ, Gu W et al (2010) Argonautes ALG-3 and ALG-4 are required for spermatogenesis-specific 26G-RNAs and thermotolerant sperm in *Caenorhabditis elegans*. Proc Natl Acad Sci U S A 107: 3588–3593
13. Burkhart KB, Guang S, Buckley BA et al (2011) Pre-mRNA-associating factor links endogenous siRNAs to chromatin regulation. PLoS Genet 7:e1002249
14. Guang S, Bochner AF, Pavelec DM et al (2008) An Argonaute transports siRNAs from the cytoplasm to the nucleus. Science 321:537–541
15. Ruby JG, Jan C, Player C et al (2006) Large-scale sequencing reveals 21U-RNAs and additional microRNAs and endogenous siRNAs in *C. elegans*. Cell 127:1193–1207
16. Claycomb JM, Batista PJ, Pang KM et al (2009) The Argonaute CSR-1 and its 22G-RNA cofactors are required for holocentric chromosome segregation. Cell 139:123–134
17. Gu W, Shirayama M, Conte D Jr et al (2009) Distinct argonaute-mediated 22G-RNA pathways direct genome surveillance in the *C. elegans* germline. Mol Cell 36:231–244
18. Hall SE, Chirn GW, Lau NC et al (2013) RNAi pathways contribute to developmental history-dependent phenotypic plasticity in *C. elegans*. RNA 19:306–319
19. Lau NC, Lim LP, Weinstein EG et al (2001) An abundant class of tiny RNAs with probable regulatory roles in *Caenorhabditis elegans*. Science 294:858–862
20. Ghildiyal M, Seitz H, Horwich MD et al (2008) Endogenous siRNAs derived from transposons and mRNAs in *Drosophila* somatic cells. Science 320:1077–1081

Chapter 7

Assays for Direct and Indirect Effects of *C. elegans* Endo-siRNAs

Philip K. Shiu, Jimmy J. Zhuang, and Craig P. Hunter

Abstract

Ever since the discovery of the first microRNAs in *C. elegans*, increasing numbers of endogenous small RNAs have been discovered. Endogenous siRNAs (endo-siRNAs) have emerged in the last few years as a largely independent class of small RNAs that regulate endogenous gene expression, with mechanisms distinct from those of piRNAs and miRNAs. Quantification of these small RNAs and their effect on target RNAs is a powerful tool for the analysis of RNAi; however, detection of small RNAs can be difficult due to their small size and relatively low abundance. Here, we describe the novel FirePlex assay for directly detecting endo-siRNA levels in bulk, as well as an optimized qPCR method for detecting the effect of endo-siRNAs on gene targets. Intriguingly, the loss of endo-siRNAs frequently results in enhanced experimental RNAi. Thus, we also present an optimized method to assess the indirect impact of endo-siRNAs on experimental RNAi efficiency.

Key words siRNA quantification, Endo-siRNA targets, RNAi efficacy, FirePlex assay, qPCR

1 Introduction

Since the first attempt by Bartel, Ambros, and colleagues to systematically clone microRNAs in *C. elegans* [1, 2], the definition of endogenous short-interfering RNAs has undergone many changes [3, 4]. In recent years, Mello, Fire, Ruvkun, and colleagues have defined *C. elegans* endo-siRNAs to refer to 22mers and 26mers that predominantly begin with a G residue. Endo-siRNAs are potentially amplified and developmentally regulated and silence endogenous gene targets with dependence on *dicer*-interacting genes [5–7]. Whereas the term "endo-siRNAs" is well defined in *C. elegans*, it is used interchangeably with or as an abbreviation for various species of endogenous small RNAs in *Drosophila* and mammals.

The cloning and sequencing of individual endo-siRNAs have been optimized extensively by Mello and colleagues, with

Andreas Werner (ed.), *Animal Endo-siRNAs: Methods and Protocols*, Methods in Molecular Biology, vol. 1173,
DOI 10.1007/978-1-4939-0931-5_7, © Springer Science+Business Media New York 2014

robust updates in response to continuing increases in sequencing technologies [8]. However, the genetic context and the regulatory impact of many of these endo-siRNAs are still mysterious despite the many large datasets available. For instance, Ruvkun and colleagues discovered that some endo-siRNAs target duplicated genes [9], while Kennedy and colleagues found that some endo-siRNAs have a nuclear preference [10]. Hence, additional higher-order analysis is needed to unravel the extensive biological functions of endo-siRNAs.

Publicly available sequencing datasets report endo-siRNAs number from hundreds to hundreds of thousands [5, 6, 11, 12]. Because of their differential sizes and termini [13–15], confounded by an amplification processing step whereby the rarer 26Gs get amplified to become the abundant 22Gs [6], understanding the role of particular endo-siRNAs in *C. elegans* can be difficult. Even more challenging, the loss of endo-siRNAs in many instances induces an increase in sensitivity to experimental RNAi [16, 17]. Therefore, it is important to assay both the direct and indirect effects of endo-siRNAs in order to understand their biological significance.

Since *C. elegans* endo-siRNAs impact on endogenous gene expression and the efficacy of experimental RNAi, we briefly survey the genetic resources available and outline specific methods to detect some of these effects. Here, we specifically examine the novel FirePlex method to detect endo-siRNA quantities in bulk. Additionally, we outline an optimized method for assaying validated endo-siRNA targets. Finally, we present our method for detecting subtle changes in experimental RNAi efficacy induced by the loss of endo-siRNAs.

1.1 Quantification of Bulk siRNAs

Quantification of siRNA levels is useful for the analysis of endo-siRNA pathways. Due to their small size, efficient detection of siRNAs by conventional methods, such as microarrays or qPCR, is difficult. In *C. elegans,* the gold standard for measuring siRNA levels is small RNA sequencing. However, this approach is time-consuming, is relatively expensive, and requires technical expertise and optimization.

Here, we introduce and describe the use of the FirePlex assay for the quantification of siRNAs in *C. elegans.* Although the FirePlex assay was originally commercialized for miRNAs, we have found that it can robustly quantify siRNA levels as well. The FirePlex platform utilizes encoded hydrogel particles to perform multiplexed detection of up to 68 targets in each well of a standard 96-well filter plate. Particles bear unique barcodes that correspond to a single target detected on each. The assay is performed in three steps—hybridization, labeling, and reporting, with rinses between each step (Fig. 1). During hybridization, targets bind to siRNA-specific DNA probes embedded in the hydrogel particles. Labeling is

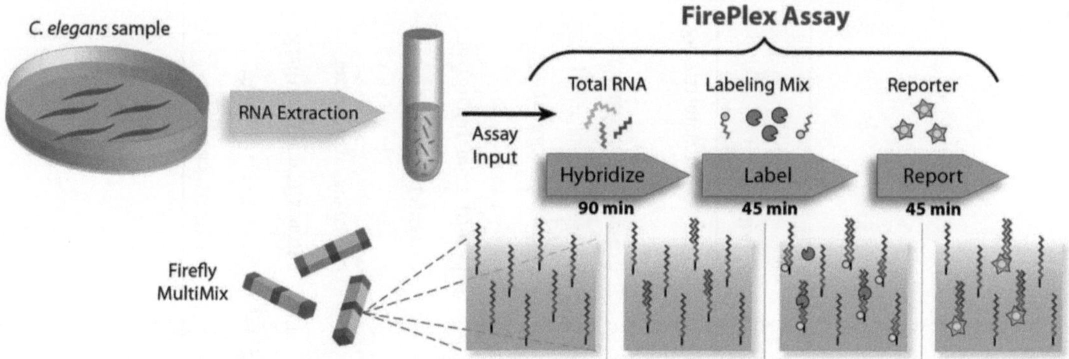

Fig. 1 Schematic outline of the FirePlex assay

accomplished via ligation of a biotinylated universal adaptor using the probe as a template. In the final step, a streptavidin-conjugated fluorescent reporter is added to visualize the hybridization event. The assay provides quantitative results, with the level of fluorescence on each particle corresponding to the amount of siRNA target present in the sample. The encoded particles are then scanned in a standard flow cytometer.

We used the FirePlex assay with purified total RNA as the input, though the system may also be applied to crude cell and tissue digests. Samples can be analyzed on-site with a FirePlex kit and a conventional flow cytometer, or alternatively, sent directly to Firefly BioWorks for custom analysis with additional costs. Although the FirePlex assay is considerably simpler than conventional siRNA sequencing, it has important limitations. First, the assay is not a discovery tool—target sequences must be known and specified. Furthermore, while the FirePlex assay allows examination of up to 68 targets per well, short RNA sequencing provides a comprehensive analysis of all siRNAs in the sample. And last, the FirePlex platform has limited sensitivity; we have been able to detect sequences found at a comparable rate of 150 reads/million in a previously published siRNA sequencing dataset (Table 1) [18]. However, our previous results suggest that the FirePlex assay is likely an order of magnitude more sensitive. Despite these disadvantages, the simplicity and throughput of the FirePlex assay should make it a very attractive tool for researchers. We imagine that the FirePlex assay can be used to quickly screen through various mutants or conditions, with interesting results being followed up with siRNA sequencing, if necessary.

1.2 Direct Quantification of Endo-siRNA Target Expression

Robust amplification is a unique aspect of RNAi in *C. elegans*, which is absent in other organisms such as mammals [19, 20]. A possible mechanistic explanation suggests that amplification plays a significant role in *C. elegans* RNAi because it allows siRNAs

Table 1
Robust endo-siRNA targets

Gene	Endo-siRNA loss via	Changes upon endo-siRNA loss	RT primer	F primer	R primer
F07G6.6	*rrf-3* and *eri-1* mutants	Increase	ctcaaaaggtcctctctcatttg	cgcacaaaacattaaatttgctaac	ccccgacaagtcaatatttttgag
F14F7.5	*rrf-3* and *eri-1* mutants	Increase	caagtttatccattgttcgtacttc	atggttttccgtgaatctgaag	gttgaagtgctctggattttaac
F39E9.7	*rrf-3* and *eri-1* mutants	Increase	caagcgcctatacttagcgatgg	gattgatgtcctgaacccagtg	gagaattgcttcggcagctg
H16D19.4	*rrf-3* mutants	Increase	ctccgattacatcttaagtgtgtag	gattacgatcaacgcaagatac	gcaatttccttatttgaatgttgc
Y37E11B.2	*rrf-3* and *eri-1* mutants	Increase	ggggtaaggtttcagcgaag	gtgcctgcttctcctcatc	cattgatcttggttgcaggttc
Y43F8B.9	*rrf-3* mutants	Increase	cagattgattgccatttcaagg	ccaccgctagctgtaaaaaatac	cactcaaagcacattggcag
C54G7.3	*rrf-3* mutants	Decrease	ggctcacttgctgagacac	gatcgtcggtcccggc	cattacttgcacaggttgtgatg
F55A4.4	*rrf-3* mutants	Decrease	gtcgggcttgcttagcgatg	aagtgcaagggagaacagaaag	gcacattggcagcaaggg
ZK816.5	*rrf-3* mutants	Decrease	gggattcggccgtcaatatc	gggtggaataatgtacttgttcac	attttgttttccgtcgcgc

to target more than one transcript through the process of transitive RNAi [21, 22]. However, when endo-siRNAs target more than one transcript, or one transcript is targeted by more than one endo-siRNAs, the relationship is between endo-siRNA presence and target gene expression often remains unclear. Generally, endo-siRNAs are thought to silence their complementary gene targets [5], but there are cases in which endo-siRNAs fail to silence or—in contrast—even protect their complementary gene targets [11].

Therefore, to monitor the direct effects of endo-siRNAs on endogenous gene expression, we have validated nine exemplary endo-siRNA targets by qPCR (Table 2). The relevant sequences derive from previously published small RNA sequencing data [22, 23]. The targets are listed below and the changes in expression upon endo-siRNA depletion are indicated. The reported targets can be used as reference genes to monitor the impact of induced changes in endo-siRNA abundance. In a related series of experiments, we describe an optimized qPCR protocol to quantitate those changes in gene expression.

1.3 Gauging Endo-siRNAs' Effects on Experimental RNAi

The canonical class of *C. elegans* endo-siRNAs was first discovered in the context of experimental RNAi. Ruvkun and colleagues performed a genetic screen for mutants with enhanced sensitivity to experimental RNAi [24]. These *e*nhanced *RNAi* (Eri) mutants were later discovered to be missing a class of small RNAs that became the founding members of what are now called *C. elegans* endo-siRNAs [5, 6, 25]. A comprehensive list of these *eri* genes and their functions has been previously described [19]. Moreover, Ruvkun and colleagues recently reported that genes responsible for the transcriptional regulation of *eri* genes were themselves susceptible to experimental RNAi with a potential impact on endo-siRNAs [26]. In addition, there are many lesser-understood genes mediating enhanced RNAi presumably also affecting endo-siRNAs. Two recent reports identified the nuclear Argonaute *nrde-3* as an essential component for nuclear exogenous and endogenous RNAi [22]. Moreover, the perinuclear foci gene *mut-16* was found to be required for the germ-line component of RNAi [27]; notably, both mutants are partially RNAi-defective.

This complex layer of gene regulation by endo-siRNAs is difficult to assess, especially when attempting to attribute direct causation. However, in most instances, perturbing the regulation of endo-siRNAs results in hyper-sensitizing or de-sensitizing experimental RNAi. Therefore, we describe here an optimized method to gauge the effect of endo-siRNAs on experimental RNAi to characterize indirect effects of endogenous gene regulation.

Note that our protocol presumes two points: First, experimental RNAi is in competition with endogenous RNAi, a well-accepted model in the field [28]; second, experimental RNAi is dosage-dependent [29]. Consequently, we assume that perturbation of endogenous RNAi affects experimental RNAi efficiency. Presented below is a protocol for detecting the differential efficiencies in experimental RNAi, described in detail from a previous study [29].

2 Materials

Always use RNase-free materials and ultrapure or DEPC-treated water.

RNA extraction and FirePlex assay:

1. 1.5 mL Phase-Lock Gel tubes, heavy (Eppendorf).
2. TRIzol Reagent (Life Technologies).
3. PTFE (Polytetrafluoroethylene) Tissue Grinder Douncer, 2 mL, glass vessel and serrated plunger (VWR).
4. Chloroform.
5. 5 M NaCl: dissolve 29.22 g of NaCl in 80 mL of water and fill up to 100 mL. Sterilize by autoclaving or sterile filter. Also available from commercial sources.
6. 20 mg/mL glycogen as a carrier for RNA precipitation. Available from commercial sources.
7. Isopropanol.
8. FirePlex kit (Firefly BioWorks, Inc., Cambridge, MA, USA).
9. A standard flow cytometer (established settings exist for Millipore Guava easyCyte 8HT, BD Accuri C6, Millipore Guava easyCyte 6HT, BD LSRFortessa, and Life Technologies Attune).
10. Vacuum Manifold for 96-well filter plates (one optimized for FirePlex is available from Firefly BioWorks).

Direct quantification of endo-siRNA target expression:

1. RNase-free recombinant DNase I (Roche Applied Science).
2. Qiagen RNeasy mini kit (Qiagen).
3. ThermoScript™ RT-PCR System for First-Strand cDNA Synthesis (Invitrogen).
4. Qiagen QuantiTect SYBR Green PCR Kit (Qiagen).
5. Twin.tec 96-well real-time PCR Plates, skirted, blue including Masterclear™ real-time PCR self-adhesive film, (Eppendorf).
6. Mastercycler® ep realplex PCR machine (Eppendorf).
7. Normal PCR thermocycler.
8. PCR plate spinner.

Table 2
Exemplary Ahringer library targets for detecting changes in RNAi sensitivity

Target	Phenotype	Strength of RNAi[a]
dpy-13	Dumpy	Weak
unc-73	Curled dumpy	Weak
lir-1	Lethal or larval arrest	Medium
dpy-11	Dumpy	Strong
pos-1	Embryonic lethal	Strong
unc-22	Twitching	Strong

[a]A "weak" RNAi target indicates that the bacterial colony containing that vector from the Ahringer library only induces a RNAi knockdown in an enhanced RNAi strain, whereas a "strong" RNAi target means that only a RNAi-attenuated or -defective strain would not exhibit RNAi knockdown

Gauging endo-siRNAs' effects on experimental RNAi:

1. Ahringer library of *C. elegans* feeding RNAi clones [30, 31, *see* Table 2] (Source BioScience, LifeSciences).
2. QIAprep Spin Miniprep Kit (Qiagen).
3. Isopropyl-β-D-thiogalactopyranoside (IPTG).
4. 35 × 10 mm plates (Greiner Bio-One).
5. Carbenicillin.
6. Tetracycline.
7. NG media components: 3 g NaCl, 2.5 Bacto Peptone (Difco) and 17 g Bacto-agar (Difco) are dissolved in 975 mL distilled water. After autoclaving, 1 mL cholesterol in ethanol (5 mg/mL), 1 mL of 1 M $CaCl_2$, 1 mL of 1 M $MgSO_4$, and 25 mL of 1 M potassium phosphate buffer (pH 6.0) are added [32].
8. LB plates (Table 3).

3 Methods

3.1 RNA Extraction (See Note 1)

Wash several (2–5) 10-cm plates of animals into a 15 mL conical centrifuge tube using water (*see* **Note 2**). Centrifuge at 11,000 × *g*, 1 min.

1. Wash three times with 15 mL of water.
2. Discard the final wash. Using a Pasteur pipette, transfer 200–350 μL of worms into a 1.5 mL microfuge tube. Freeze at −80 °C for 1 h.

3. Prepare phase-lock tubes by centrifugation at 11,000 × *g* for 1 min.
4. Add 400 μL TRIzol to each sample.
5. Vortex samples at room temperature until the TRIzol–worm mixture is a slurry.
6. Transfer the solution into a douncer on ice using an RNase-free glass Pasteur pipette.
7. Dounce slurry with a twisting motion 20 times on ice. Between samples clean with RNaseOUT and DEPC-treated water or use a new douncer (*see* **Note 3**).
8. Transfer the lysed worms to a phase-lock tube.
9. Add 80 μL chloroform, mix by inversion for 1 min, and incubate at room temperature for 5 min.
10. Centrifuge at 15,000 × *g* for 15 min at 4 °C.
11. Transfer the top aqueous layer to a siliconized 1.5 mL microfuge tube.
12. Add, in this order, 50 μL of 5 M NaCl, 2.5 μL of 20 mg/mL glycogen, and 800 μL of isopropanol to each tube.
13. Mix by inversion for 1 min and place at −80 °C for 1 h.
14. Centrifuge at 11,000 × *g* for 15 min at 4 °C.
15. Remove the supernatant with a pipette.
16. Wash with 500 μL of ice-cold 100 % ethanol.
17. Spin at 11,000 × *g* at 4 °C for 1 min.
18. Remove the supernatant and let remaining ethanol evaporate for about 5 min at room temperature. Do not allow the pellet to dry out completely.
19. Dissolve the RNA pellet in 80 μL of water by vortexing at room temperature.
20. Quantitate the RNA concentration using a NanoDrop or an equivalent spectrophotometer.

At this point, samples can be sent directly to Firefly Bio for commercial analysis. Alternatively, samples can be processed in house using a standard flow-cytometer.

3.2 FirePlex Assay

1. Cut the plate seal to expose assay wells on the filter plate required for the experiment (provided in the assay kit *see* **Note 4**).
2. Dilute sample to twice the final concentration; in our hands this would routinely be 200 ng/μL (*see* **Note 5**).
3. Invert, then vortex MultiMix (provided with the assay) for 3 s. Add 35 μL of MultiMix to each well. Mix by pipetting up and down (*see* **Note 6**).
4. Vacuum-filter the plate and wipe the bottom dry (*see* **Notes 7–9**).

5. Add 25 μL of hybridization buffer to each well, followed by 25 μL of sample.
6. Hybridize the plate for 90 min at 37 °C shaking at 750 rpm (*see* **Note 10**).
7. Near the end of the hybridization step, prepare 1× rinse buffer by mixing 0.2 mL of 10× Rinse Buffer with 1.8 mL of water per assay well. Prepare the Labeling Mix by mixing 78 μL of water, 1.6 μL of 50× labeling buffer, and 0.4 μL of labeling enzyme per assay well.
8. After hybridization, add 200 μL 1× Rinse Buffer to each well and vacuum-filter the plate.
9. Repeat **step 8**. Blot dry the bottom side of the plate.
10. Add 75 μL of Labeling Mix to each well and shake at room temperature at 750 rpm for 45 min.
11. Prepare the Reporter Mix by mixing 64 μL of water with 16 μL of 5× Reporter Solution per each assay well.
12. After Labeling, add 200 μL 1× Rinse Buffer to each well and vacuum-filter the plate.
13. Repeat **step 13**. Blot dry the bottom side of the plate.
14. Add 75 μL Reporter Mix to each well. Shake at room temperature at 750 rpm for 45 min. After the incubation, add 200 μL 1× Rinse Buffer to each well and vacuum-filter the plate.
15. Repeat **step 17**. Blot dry the bottom side of the plate.
16. Add 175 μL Run Buffer to each well.
17. Apply the correct scan settings for the specific flow-cytometer used (provided at http://www.fireflybio.com/productsupport).
18. Scan at least 100 μL of sample for each well.
19. Save the FCS file for analysis using the FireCode software (http://www.fireflybio.com/productsupport) (Table 3).

3.3 Direct Quantification of Endo-siRNA Target Expression

1. To remove traces of genomic DNA from the extracted RNA, combine the following mixture in a PCR tube: 5 μL 10× DNase I buffer, 4 μL DNase I, and 40 μL of the RNA sample. Incubate the sample in the PCR machine at 37 °C for 20 min and denature the enzyme at 75 °C for 20 min (*see* **Note 13**).
2. Isolate the RNA using the Qiagen RNeasy kit. Transfer the reaction mix to a fresh eppendorf tube and add 50 μL of water and 350 μL of buffer RLT.
3. Add 250 μL of ethanol, mix, and immediately transfer to spin column (*see* **Note 11**).
4. Centrifuge at 11,000 ×*g* for 15 s and discard the flow-through carefully.

Table 3
The FirePlex assay is consistent with siRNA sequencing

Genotype	Relative pos-1 siRNA levels (FirePlex assay)	Relative pos-1 single-stranded siRNA levels (predicted from literature)
Wild-type	1	1
rde-10	0.109 ± 0.0148	0.11 (Yang et al. [33]; Zhang et al. [18])
rde-11	0.125 ± 0.0072	0.12 (Yang et al. [33])
rde-1	0.022 ± 0.024	~0 (Steiner et al. 2009 [34])

Animals were exposed to *pos-1* dsRNA and *pos-1* siRNA levels were measured

5. Wash the column with 500 μL of buffer RPE, centrifuge at 11,000 × *g* for 15 s, and discard the flow-through carefully.
6. Wash the column again with 500 μL buffer RPE, centrifuge at 11,000 × *g* for 2 min, and discard the flow-through carefully.
7. Put the column into a new collection tube, centrifuge at 11,000 × *g* for 1 min, and discard flow-through.
8. Put the column in a 1.5 mL collection tube, add 30 μL of water, and elute the RNA by centrifugation at 11,000 × *g* for 1 min.
9. With the column still in the same collection tube, add another 30 μL of water and centrifuge at 11,000 × *g* for 1 min resulting in slightly less than 60 μL of RNA containing eluate. Spec the eluted cleaned RNA on the NanoDrop.
10. Determine the concentration of the RNA by NanoDrop or an equivalent spectrophotometer. Dilute the samples to a concentration of about 200 ng/μL. These samples should be stored at −80 °C as a working stock; repeated freeze–thaw cycles should be avoided.
11. The reverse transcription (RT) reaction is assembled in a PCR tube and contains 1 μL 10 μM RT primer, 3 μL of total RNA, 2 μL of 10 mM dNTP mix, and 6 μL of water (Invitrogen Thermoscript kit) (*see* **Note 12**).
12. Incubate in PCR machine at 65 °C for 5 min, transfer the tube on ice.
13. Meanwhile, assemble a master mix containing per reaction, 4 μL of 5× cDNA synthesis buffer, 1 μL of 0.1 M DTT, 1 μL of RNaseOUT (RNAse inhibitor), 1 μL of water and 1 μL of Thermoscript reverse transcriptase (Invitrogen Thermoscript kit). Add 8 μL of master mix to each tube containing RNA, primer and dNTPs, mix well.
14. Incubate the RT reaction in a PCR machine at 50 °C for 1 h and then denature the enzyme at 85 °C for 5 min. Transfer the reaction on ice.

15. Add 1 μL of RNase H (Invitrogen Thermoscript kit) to each tube and incubate at 37 °C for 20 min. Transfer the reaction on ice. The cDNA can be stored at −20 °C for months.
16. Thaw Qiagen 2× QuantiTect SYBR Green PCR master mix, RNase-free water, cDNA, and primers and keep the vials on ice (QuantiTect SYBR Green PCR Kits).
17. Per Eppendorf skirted qPCR plate well, add in the following order: 20 μL of RNase-free water, 1.5 μL of forward primer (10 μM), 1.5 μL of reverse primer (10 μM), 2 μL of cDNA, and 25 μL of SYBR green master mix. Each cDNA-primer-pair combination should be done in technical triplicates (*see* **Note 14**).
18. Mix the contents well and seal plate with clear Eppendorf tape using a tape roller. Spin the reaction down in a PCR plate spinner.
19. Run the reaction on a thermocycler that allows for quantitative monitoring of the reaction. We use a Mastercycler® ep realplex PCR machine with BOTH dye and probe as SYBR green settings and the following cycling conditions: Activation at 95 °C for 15 min, 45 cycles of 94 °C for 15 s, 52 °C for 30 s, 72 °C for 1 min, followed by a melting curve analysis. Record all C_t values after Noise band adjustment.
20. Relative transcript levels are then quantified using the $-\Delta\Delta C_t$ method [35] (*see* **Note 15**).

3.4 Gauging Endo-siRNAs' Effects on Experimental RNAi

1. For every liter of medium, add 17 g of Bacto agar, 2.5 g of Bacto Peptone, and 3 g of sodium chloride. Add in stirring bar and autoclave. Stir the solution after autoclaving until you can touch the flask. Add 1 mL of 5 mg/mL cholesterol in ethanol, 1 mL of 1 M $CaCl_2$, 1 mL of 1 M $MgSO_4$, 25 mL of 1 M KH_2PhO_4, 1 mL of 100 g/l carbenicillin, and 238 mg of IPTG while stirring. Pour a volume of 3.5 mL media per 35 × 10 mm plate. Keep the plates at room temperature overnight; then they can be stored at 4 °C for up to 2 weeks.
2. Add and spread the antibiotics to LB plates, carbenicillin, to a final concentration 100 mg/L, and tetracycline to a final concentration of 12.5 mg/L. Leave to dry at room temperature. The plates can be stored in 4 °C for up to 2 weeks.
3. From the Ahringer library, seek the RNAi vector colony against your gene of interest via the Source BioScience LifeSciences database(http://www.lifesciences.sourcebioscience.com/clone-products/mirn--nai-resources/c-elegans-rnai-library.aspx) (*see* **Note 16**).
4. Take an autoclaved toothpick, and scrape a bit of bacteria carrying the vector from the desired library well, and streak in thirds onto a prewarmed LB-carb/tet plate. Grow at 37 °C overnight.

5. The next morning, label 5–10 individual colonies to be verified for correct plasmid presence.
6. Prepare 2 mL of LB of medium in a culture tube for each colony and add 2 μL of 100 g/L carbenicillin and 5 μL of 5 mg/mL tetracycline.
7. Use an autoclaved toothpick, to inoculate the culture, label both plate and culture accordingly. Grow the bacteria at 37 °C in a shaker for no longer than 15 h. Label each tube with the colony number. Isolate the plasmids. In our hands, the QIAprep Spin Miniprep Kit along with the provided protocol works well. Check the insert by sequencing using the M13 forward primer [31].
8. Regrow the verified colonies, either scale volume or flask size up about five times or grow 3–5 liquid culture tubes per RNAi gene target. Always include a colony containing the empty vector L4440 as a negative control for the knock-down experiment; grow as many liquid cultures of the negative control strain as there are RNAi-positive cultures.
9. After overnight incubation, take an aliquot of each culture and dilute it 1/10× with LB media.
10. Measure absorbance at 600 nm in the spectrometer for each culture. Based from this measurement, prepare an array of at least five different optical densities of bacteria representing different potencies of RNAi-induction (e.g., have cultures of OD_{600nm} of 0.25, 0.5, 1.0, 1.5, and 2.0) (*see* **Note 17**).
11. A culture with a final OD_{600nm} lower than 1.0 should be diluted with vector L4440-containing bacteria, and a culture with a final OD_{600nm} higher than 1.0 should be diluted with just LB media (*see* **Note 3**).
12. Add 50 μL of each culture of the array (RNAi gene target × dilution series of bacteria) to the middle of IPTG-carb plates (Subheading 3.4, **step 1**) (*see* **Note 18**).
13. Leave at room temperature for 1 day to allow for IPTG induction of the T7 promoters to drive dsRNA expression.
14. Place non-starved third-larval stage single *C. elegans* worms of the desired strain onto each dosage of an RNAi-bacteria containing plate, perform in triplicates (*see* **Notes 19** and **20**).
15. Allow worms to hatch at the desired temperature (e.g., 20 °C) until the next generation reaches young adulthood.
16. Score for penetrance of the RNAi knockdown phenotype on each plate, and calculate the average and standard deviation for each dosage (*see* **Note 20**).

As noted before, this protocol only gives a direction in which experimental RNAi is affected, which in most cases suggests that endo-siRNAs are perturbed. However, the directional relationship between an increase or decrease of endo-siRNAs and experimental RNAi is not always the inverse [27, 36]. Therefore, this protocol is appropriate only for indirectly measuring the likely presence or absence of perturbations to endo-siRNA efficacy, and not the direction of the change in efficacy.

4 Notes

1. Standard procedures to avoid contamination with RNase should be used: wear gloves at all times, use filter tips, and spray work surfaces and instruments with RNaseOUT. Total RNA extraction can also be performed by any other standard method.
2. The typical yield from this extraction exceeds 1 mg of RNA, while we typically require 5 μg per FirePlex reaction. Therefore, the number of starting worms can be scaled down.
3. To avoid the cleaning of the douncer between samples, it is convenient to have multiple douncers.
4. The assay plate is purchased from Firefly BioWorks. It is useful to include a small RNA control that will not change between samples to normalize RNA input, for example, U18.
5. We use a final concentration of 100 ng/μL (5 μg total). This amount can be changed depending on the abundance of your target.
6. Ensure that particles are mixed well by pipetting up and down between each well.
7. For all filtration steps, maintain vacuum pressure below 2 PSI during filtration. Do not over-filter particles. Bubbles are OK. While performing the assay, cover the plate, but do not use a plate seal.
8. After all filtration steps, resuspend particles within 30 s of filtering. It may be convenient to use a multichannel pipette if many samples are run in parallel.
9. Ensure that the bottom of the plate is always dry.
10. The optimal shaking speed depends on the orbit of the shaker used. The speed of 750 rpm is best suited for an orbit of 3 mm.
11. For **steps 2–10**, follow the protocol provided by Qiagen.
12. Perform normal PCR with cDNA and RNA-alone with F&R primers for quality control: RNA-alone should get no bands and cDNA's amplified bands should correspond to mRNA length and not DNA lengths.
13. The DNase treatment of the RNA and the RT can be assessed by end-point PCR using forward and reverse primers with

either cDNA or RNA as a template. Whereas amplification of cDNA should result in the expected amplified fragment the RNA should not yield any product.

14. Depending on the thermocycler and the reaction plate the assay volume can be reduced to 25 μL or even 10 μL.
15. While there are many ways to quantitate expression levels of genes after obtaining C_t values from qPCR, we favor the $-\Delta\Delta C_t$ method because it is a relative measure. Mathematically, the method first normalizes the difference between a gene of interest against a housekeeping gene (the first Δ), then normalizes that value to the difference in a control condition/strain (the second Δ) [35]. In our hands this is an appropriate method for studying gene regulation by endo-siRNAs because endo-siRNA levels may impact other aspects of *C. elegans* biology [23]. Therefore, it is important that the expression of the gene of interest is normalized to a dynamic housekeeping marker which may experience some of those same hereto unknown influences of endo-siRNAs as well. Logistically speaking, it means each sample must be accompanied by an RT reaction against a housekeeping. In our experience, *gpd-3*, *ama-1*, and *pmp-3* are quite robust and stable in expression throughout all worm developmental stages. For more sophisticated temporal normalizations, we found *dpy-13* to be quite useful for measuring genes expressed near the first three larval molts, *bli-1* useful for late L4 genes, *vit-2* for early young adult genes, and *mex-3* and *pgl-1* for germ-line genes. In the Source BioScience LifeSciences database, the library against which there are RNAi clones is listed by their gene sequence number listed on WormBase. After searching with the gene sequence number, examine the listed forward and reverse primers used to build the plasmid portion between T7 promoters driving the dsRNA, to ensure that the region covers an exonic fragment of your gene of interest.
16. The optical density of the bacteria carrying the RNAi-vector is used as a proxy for RNAi dosage, as previously described in two independent studies [29, 37]. A robust dilution series of RNAi dosages causes a graded RNAi phenotypic response and potentially allows to differentiate subtle sensitivities. In our experience, RNAi targets vary tremendously in their sensitivity. For example, *unc-22 (RNAi target)* is extremely sensitive, and bacteria at an OD_{600nm} of 1/100,000 still produce some twitching animals. Conversely, *dpy-13(RNAi target)* is extremely insensitive, and OD_{600nm} of 4.5 sometimes still does not induce Dpy phenotypic animals. Therefore, it is important to examine a wide enough range of RNAi dosages to ensure that a robust response gradient is available for comparison of sensitivity among strains. Ideally, OD_{600nm} concentrations series ought to

include concentrations which induce no and fully penetrant RNAi knockdown.

17. In our experience, a minimal total bacteria volume of 50 μL at a concentration of 1.0 OD_{600nm} is necessary to ensure that the progeny of a single worm do not starve when scoring occurs.
18. Triplicates of triplicates provide the most robust data for dosage response curve analysis. Therefore, ensure there are enough cultures grown to accommodate the different replicates and dosages. 15 h of growth at 37 °C usually generates around an OD_{600nm} of 2.0 in each 2 mL liquid culture.
19. For most phenotypes, starting the knockdown in the mother and scoring the RNAi phenotype in the offspring is the most robust and convenient method of RNAi. At the molecular level, there is both intra- and inter-generational RNAi occurring in this experimental set up [38, 39]. Furthermore, for germline RNAi targets, starting the knockdown in worms at stage L3 allows enough time for at least some level of knockdown to be observed. Under these circumstances, the reduction in brood size is scored as a read-out for RNA interference. However, gonadal RNAi targets require first-larval stage single worms to be placed to ensure the elimination of the gonad.
20. RNAi sensitivity is expected to result in a sigmoidal response curve which varies between strains and conditions if enough dosages are measured. The greatest variability should occur in the intermediate dosages. If the RNAi knockdown phenotype does not vary significantly across the dilution series, a different range of RNAi-bacteria concentration dosages should be assessed.

References

1. Lau NC, Lim LP, Weinstein EG, Bartel DP (2001) An abundant class of tiny RNAs with probable regulatory roles in Caenorhabditis elegans. Science 294:858–862
2. Ambros V, Lee RC, Lavanway A, Williams PT, Jewell D (2003) MicroRNAs and other tiny endogenous RNAs in C-elegans. Curr Biol 13:807–818
3. Ruby JG, Jan C, Player C, Axtell MJ, Lee W, Nusbaum C, Ge H, Bartel DP (2006) Large-scale sequencing reveals 21U-RNAs and additional microRNAs and endogenous siRNAs in C. elegans. Cell 127:1193–1207
4. Lee RC, Hammell CM, Ambros V (2006) Interacting endogenous and exogenous RNAi pathways in Caenorhabditis elegans. RNA 12: 589–597
5. Vasale JJ, Gu W, Thivierge C, Batista PJ, Claycomb JM, Youngman EM, Duchaine TF, Mello CC, Conte D Jr (2010) Sequential rounds of RNA-dependent RNA transcription drive endogenous small-RNA biogenesis in the ERGO-1/Argonaute pathway. Proc Natl Acad Sci U S A 107:3582–3587
6. Gent JI, Lamm AT, Pavelec DM, Maniar JM, Parameswaran P, Tao L, Kennedy S, Fire AZ (2010) Distinct phases of siRNA synthesis in an endogenous RNAi pathway in C. elegans soma. Mol Cell 37:679–689
7. Zhang C, Montgomery TA, Gabel HW, Fischer SEJ, Phillips CM, Fahlgren N, Sullivan CM, Carrington JC, Ruvkun G (2011) mut-16 and other mutator class genes modulate 22G and 26G siRNA pathways in Caenorhabditis elegans. Proc Natl Acad Sci U S A 108:1201–1208
8. Gu WF, Lee HC, Chaves D, Youngman EM, Pazour GJ, Conte D, Mello CC (2012) CapSeq and CIP-TAP identify Pol II start sites and

reveal capped small RNAs as C. elegans piRNA precursors. Cell 151:1488–1500

9. Fischer SE (2010) Small RNA-mediated gene silencing pathways in C. elegans. Int J Biochem Cell Biol 42:1306–1315
10. Guang S, Bochner AF, Burkhart KB, Burton N, Pavelec DM, Kennedy S (2010) Small regulatory RNAs inhibit RNA polymerase II during the elongation phase of transcription. Nature 465:1097–1101
11. Claycomb JM, Batista PJ, Pang KM, Gu W, Vasale JJ, van Wolfswinkel JC, Chaves DA, Shirayama M, Mitani S, Ketting RF, Conte D Jr, Mello CC (2009) The Argonaute CSR-1 and its 22G-RNA cofactors are required for holocentric chromosome segregation. Cell 139: 123–134
12. Conine CC, Batista PJ, Gu W, Claycomb JM, Chaves DA, Shirayama M, Mello CC (2010) Argonautes ALG-3 and ALG-4 are required for spermatogenesis-specific 26G-RNAs and thermotolerant sperm in Caenorhabditis elegans. Proc Natl Acad Sci U S A 107:3588–3593
13. Kamminga LM, van Wolfswinkel JC, Luteijn MJ, Kaaij LJ, Bagijn MP, Sapetschnig A, Miska EA, Berezikov E, Ketting RF (2012) Differential impact of the HEN1 homolog HENN-1 on 21U and 26G RNAs in the germline of Caenorhabditis elegans. PLoS Genet 8:e1002702
14. Montgomery TA, Rim YS, Zhang C, Dowen RH, Phillips CM, Fischer SE, Ruvkun G (2012) PIWI associated siRNAs and piRNAs specifically require the Caenorhabditis elegans HEN1 ortholog henn-1. PLoS Genet 8: e1002616
15. Billi AC, Alessi AF, Khivansara V, Han T, Freeberg M, Mitani S, Kim JK (2012) The Caenorhabditis elegans HEN1 ortholog, HENN-1, methylates and stabilizes select subclasses of germline small RNAs. PLoS Genet 8:e1002617
16. Gent JI, Schvarzstein M, Villeneuve AM, Gu SG, Jantsch V, Fire AZ, Baudrimont A (2009) A Caenorhabditis elegans RNA-directed RNA polymerase in sperm development and endogenous RNA interference. Genetics 183: 1297–1314
17. Pavelec DM, Lachowiec J, Duchaine TF, Smith HE, Kennedy S (2009) Requirement for the ERI/DICER complex in endogenous RNA interference and sperm development in Caenorhabditis elegans. Genetics 183:1283–1295
18. Zhang C, Montgomery TA, Fischer SEJ, Garcia SMDA, Riedel CG, Fahlgren N, Sullivan CM, Carrington JC, Ruvkun G (2012) The Caenorhabditis elegans RDE-10/RDE-11 complex regulates RNAi by promoting secondary siRNA amplification. Curr Biol 22:881–890
19. Zhuang JJ, Hunter CP (2012) RNA interference in Caenorhabditis elegans: uptake, mechanism, and regulation. Parasitology 139: 560–573
20. Pak J, Fire A (2007) Distinct populations of primary and secondary effectors during RNAi in C. elegans. Science 315:241–244
21. Alder MN, Dames S, Gaudet J, Mango SE (2003) Gene silencing in Caenorhabditis elegans by transitive RNA interference. RNA 9: 25–32
22. Zhuang JJ, Banse SA, Hunter CP (2013) The nuclear argonaute NRDE-3 contributes to transitive RNAi in Caenorhabditis elegans. Genetics 194:117–131
23. Zhuang JJ, Hunter CP (2012) The influence of competition among small RNA pathways on development. Genes 3:671–685.
24. Kennedy S, Wang D, Ruvkun G (2004) A conserved siRNA-degrading RNase negatively regulates RNA interference in C. elegans. Nature 427:645–649
25. Fischer SE, Montgomery TA, Zhang C, Fahlgren N, Breen PC, Hwang A, Sullivan CM, Carrington JC, Ruvkun G (2011) The ERI-6/7 helicase acts at the first stage of an siRNA amplification pathway that targets recent gene duplications. PLoS Genet 7:e1002369
26. Wu X, Shi Z, Cui M, Han M, Ruvkun G (2012) Repression of germline RNAi pathways in somatic cells by retinoblastoma pathway chromatin complexes. PLoS Genet 8:e1002542
27. Phillips CM, Montgomery TA, Breen PC, Ruvkun G (2012) MUT-16 promotes formation of perinuclear mutator foci required for RNA silencing in the C. elegans germline. Genes Dev 26:1433–1444
28. Duchaine TF, Wohlschlegel JA, Kennedy S, Bei Y, Conte D Jr, Pang K, Brownell DR, Harding S, Mitani S, Ruvkun G, Yates JR 3rd, Mello CC (2006) Functional proteomics reveals the biochemical niche of C. elegans DCR-1 in multiple small-RNA-mediated pathways. Cell 124: 343–354
29. Zhuang JJ, Hunter CP (2011) Tissue-specificity of Caenorhabditis elegans enhanced RNAi mutants. Genetics 188(1):235–237
30. Timmons L, Fire A (1998) Specific interference by ingested dsRNA. Nature 395:854
31. Kamath RS, Ahringer J (2003) Genome-wide RNAi screening in Caenorhabditis elegans. Methods 30:313–321
32. Brenner S (1974) The genetics of Caenorhabditis elegans. Genetics 77:71–94

33. Yang H, Zhang Y, Vallandingham J, Li H, Florens L, Mak, HY (2012) The RDE-10/ RDE-11 complex triggers RNAi-induced mRNA degradation by association with target mRNA in *C elegans*. Genes & development 26:846–856
34. Steiner FA, Okihara KL, Hoogstrate SW, Sijen T, Ketting RF (2009) RDE-1 slicer activity is required only for passenger-strand cleavage during RNAi in *C elegans*. Nat Struct Mol Biol 16:207–211
35. Livak KJ, Schmittgen TD (2001) Analysis of relative gene expression data using real-time quantitative PCR and the 2(-Delta Delta C(T)) Method. Methods 25:402–408
36. Tabach Y, Billi AC, Hayes GD, Newman MA, Zuk O, Gabel H, Kamath R, Yacoby K, Chapman B, Garcia SM, Borowsky M, Kim JK, Ruvkun G (2013) Identification of small RNA pathway genes using patterns of phylogenetic conservation and divergence. Nature 493:694–698
37. Rea SL, Ventura N, Johnson TE (2007) Relationship between mitochondrial electron transport chain dysfunction, development, and life extension in Caenorhabditis elegans. PLoS Biol 5:2312–2329
38. Burton NO, Burkhart KB, Kennedy S (2011) Nuclear RNAi maintains heritable gene silencing in Caenorhabditis elegans. Proc Natl Acad Sci U S A 108:19683–19688
39. Buckley BA, Burkhart KB, Gu SG, Spracklin G, Kershner A, Fritz H, Kimble J, Fire A, Kennedy S (2012) A nuclear Argonaute promotes multigenerational epigenetic inheritance and germline immortality. Nature 489:447–451

Chapter 8

Extraction and Nonradioactive Detection of Small RNA Molecules

Mark Carlile and Andreas Werner

Abstract

The emergence of small RNAs as key and potent regulators of gene expression has prompted the need for robust detection and assay protocols to be developed for investigating their generation and tissue distribution. The physicochemical nature of these RNAs allows traditional assay methods to be employed; however, due to the relatively small size of endo-siRNAs, key changes to these protocols are required. Here, we present a method for the nonradioactive detection of endo-siRNAs in mouse tissue and microinjected *Xenopus* oocytes. The method comprises steps for RNA extraction, PAGE, and low-stringency northern blotting using DIG-labelled RNA probes. Moreover, it includes a strategy to design and generate cheap hybridization probes with greatly increased sensitivity. These methods may be used as a simple and robust protocol for nonradioactive detection of small RNAs or be combined with other strategies to potentially enhance signal intensity.

Key words siRNA, Nonradioactive probe, Digoxigenin, Northern blotting endo-siRNA

1 Introduction

Functional small RNAs have been identified throughout the plant and animal kingdoms and are derived from mechanistically similar pathways for modulating gene expression. Posttranscriptional gene silencing (PTGS) in plants, RNA interference (RNAi) in animals [1], and RNA-mediated gene silencing plants [2] and lower eukaryotes, have been shown to utilize small 21–26 nt RNA molecules as functional silencing effectors.

The generation of endogenous small interfering RNA (endo-siRNA) molecules can derive from the processing of sequence-complementary mRNA molecules (natural antisense transcripts) and is carried out by RNAse-III-related enzymes [3]. Recently, siRNAs derived from the sodium phosphate co-transporter *Slc34a1* and its associated overlapping antisense transcript in murine tissues have been identified. In murine kidney, where the Na–phosphate co-transporter fulfils its physiological function endo-siRNAs

Andreas Werner (ed.), *Animal Endo-siRNAs: Methods and Protocols*, Methods in Molecular Biology, vol. 1173,
DOI 10.1007/978-1-4939-0931-5_8, © Springer Science+Business Media New York 2014

complementary to the antisense transcript were detected, whereas in testis both sense and antisense endo-siRNA orientations were found [4].

Here, we describe the modified and extended methods how the endo-siRNAs were detected in murine tissues through the use of nonradioactive detection strategies. Small RNAs are extracted and separated by polyacrylamide gel electrophoresis (PAGE), transferred to a positively charged membrane via northern blotting, and detected by hybridization of sequence complementary full-length transcript digoxigenin-labelled RNA probes. This method is a robust and sensitive assay for small RNAs; circumvents the safety issues and high-cost of using radiolabelled probes, and can serve as a starting point for the detection of small RNA from a variety of RNA sources.

2 Materials

This method was applied to RNA samples from a variety murine tissues, zebrafish embryos, and microinjected *Xenopus* oocytes. RNAse-free conditions should be maintained in all solutions and throughout all procedures.

2.1 RNA Extraction

1. Tissue homogenization: We use a small polytron mixer to homogenize the tissue in TRIzol. Alternatively, a porcelain mortar and pestle to powder the N_2-frozen tissue or plastic pestles fitting a microfuge tube can be used in the latter cases.
2. TRIzol or an equivalent chaotropic reagent.
3. Chloroform.
4. Isopropanol.
5. 75 % Ethanol in water.

2.2 Gel Electrophoresis and Northern Blotting

1. RNA loading buffer (Ambion).
2. Precast 4–12 % BisTris Polyacrylamide gels (Bio-Rad).
3. Short RNA marker (Ambion).
4. Positively charged Nylon membranes (Roche).
5. Tris-Borate-EDTA (TBE) buffer for PAGE and blotting: 89 mM Tris-base, 2 mM EDTA, 89 mM Boric acid, pH 8.5.

2.3 Probe Generation

1. In vitro RNA synthesis kit or individual components such as SP6 and T7 RNA polymerase (including buffer) and NTPs.
2. Digoxygenin-labelled NTP mix (Roche).
3. Custom-synthesized long oligonucleotide comprising T7 and SP6 adapter sequences.
4. PCR fragment purification kit (Qiagen or Sigma for example).

2.4 Hybridization and Detection

1. DIG Easy Hyb solution made up from granules (Roche).
2. DIG-labelled probe (*see* Subheading 3.4).
3. Anti-DIG detection kit.
4. X-ray film.

3 Methods

General guidance: Good laboratory practice and sterile technique are particularly applicable when working with RNA. A common reason for poor results obtained in RNA assay protocols is due to contamination by RNases, which can rapidly degrade RNA into a mixture of fragments of varying sizes. RNases affect the downstream operations and are particularly resilient to removal and/or degradation. If the following recommendations are followed, RNases should not be a concern:

- Always use sterilized water, preferably nuclease-free water from a trusted supplier; or use water treated with DEPC which reacts with amines and inactivates most enzymes. All reagents and buffers should be prepared with this water.
- The use of gloves is recommended for all operations.
- RNases are not efficiently removed by autoclaving. The use of freshly opened bags of non-sterilized plasticware (e.g., microfuge tubes) can be utilized since the manufacturing process effectively results in an RNase-free product.
- Note that degradation of the RNA sample in any of the experimental steps outlined below may only be noticeable after the final detection step.

3.1 RNA Extraction

The source of RNA (e.g., cultured cells, isolated tissue) will determine the initial steps required to isolate good-quality RNA. Minimize the length of time taken from initial sample harvest to RNA extraction initiation and keep all samples chilled or frozen at −70 °C or in liquid nitrogen if samples are stored.

For routine RNA extractions, chaotropic/phenol-based reagents are used for the initial extraction followed by precipitation with isopropanol [5]. This method has been incorporated into convenient kit form by many suppliers (e.g., TRIzol reagent) and can be used easily and effectively for a wide range of sample types. The method can also be used to extract DNA and protein from the same sample (not covered in this protocol). Most standard procedures are effective in isolating RNA in the range of 0.1–15 kb in length; however, for the extraction of small RNAs the standard incubation and centrifugation conditions need to be extended for optimal results. There are specific small RNA extraction kits available from multiple suppliers that utilize silica-based spin columns for isolating small RNAs. Column purification follows either directly after sample homogenization or, more effectively, after extraction from the organic phase. The columns are optimized for small RNA isolation by size exclusion and should not be substituted for standard nucleic acid spin columns. Due to the cooperative nature of RNA binding to these columns, processing RNA-rich samples can generate a better overall yield.

For the purposes of this protocol, a modified TRIzol extraction procedure is described that has been effective to isolate small RNAs from mouse tissue, zebrafish embryos, and microinjected *Xenopus* oocytes [6].

TRIzol reagent extraction procedure modified for extraction of small RNAs:

1. Add cold TRIzol reagent as soon as the sample is harvested. Use 1 mL of TRIzol reagent per 100 mg of tissue or 10^6 cultured cells. Depending on the tissue homogenize the sample with a microfuge pestle and/or by passage through narrow-bore injection needles (21–25G) (*see* **Note 1**). Vigorously vortex the sample for 30 s.
2. Allow the homogenate to stand unagitated at room temperature for 10 min (*see* **Note 2**).
3. Add 0.2 mL of chloroform per mL of TRIzol added in Subheading 3.1, **step 1** and vortex for 30 s.
4. Centrifuge the sample at 12,000 × *g* for 45 min at 4 °C (*see* **Note 3**).
5. Transfer the RNA-containing aqueous phase to a clean tube, add 0.5 mL of cold isopropanol per mL of TRIzol reagent, and gently mix for 2 min. Allow to stand for 20 min at −20 °C.
6. Centrifuge the sample at 12,000 × *g* for 45 min at 4 °C. Orient the tubes consistently within the rotor as to identify the putative RNA "pellet zone" in samples containing low amounts of RNA.
7. Remove the supernatant, ensuring that the pellet zone is not disturbed. It is possible that the pellet will is not visible at this point.
8. Wash the pellet with 1 mL of 75 % ethanol per mL of TRIzol. Gently mix the samples and keep them at −20 °C for 5 min. Centrifuge at 12,000 × *g* for 30 min at 4 °C, maintain the orientation of the tubes as in the previous centrifugation step.
9. Remove the supernatant and allow the pellet to air-dry for 30 min.
10. The pellet can be solubilized in nuclease-free water or a buffer compatible with downstream processing. Note that RNA at high concentrations tends to be more stable than highly diluted samples.
11. Store the extracted RNA at −70 °C.

3.2 RNA Separation

The separation of whole RNA samples in the assay of small RNAs is routinely carried out on denaturing urea polyacrylamide gels. We found that urea interfered with signal efficiency or increased background. We argue that secondary structures in short RNAs are unlikely to form during electrophoresis even under native conditions. We therefore used denaturing loading buffer and initial heat denaturation followed by conventional PAGE in TBE buffer.

Separation of RNA by polyacrylamide gel electrophoresis

Pre-run a 4–12 % Bis-Tris polyacrylamide gel in 1× TBE running buffer at 200 mV for 60 min (*see* **Notes 4** and **5**).

Mix the RNA sample with RNA loading buffer in the ratio of 1:3 (*see* **Note 6**).

Heat the sample at 75 °C for 5 min followed by brief vortexing and place the sample on ice prior to loading.

Load the RNA onto the pre-run gel and separate in fresh 1× TBE running buffer at 200 mV for 90 min or until the bromophenol blue band has migrated to the bottom of the gel.

3.3 Northern Blotting

1. Ensure that the equipment is thoroughly cleaned prior to starting the blotting procedure. If the blotting chamber is also used for western blotting, sanitization with 0.1 N NaOH for >60 min is required. Afterwards all components should be rinsed extensively and dried.
2. Remove the gel from the plastic support casing, cut off the wells and the bottom right-hand corner for orientation purposes.
3. Immerse the gel in 100 mL of fresh 1× TBE buffer and gently agitate for 10 min.
4. Pre-soak 6 Whatman 3 MM filter papers (cut to the same dimensions as the trimmed gel) and any absorptive sponges associated with the electro-blotting apparatus in 100 mL of 0.5× TBE buffer.
5. In a separate vessel soak a positively charged nylon membrane (cut to the same dimensions as the trimmed gel) in 50 mL of 0.5× TBE buffer and agitate for 10 min (*see* **Note** 7).
6. Assemble the electro-blot according to the supplier's instructions and transfer for 90 min in agitated 0.5× TBE buffer.
7. Disassemble the electro-blot, remove the membrane using tweezers and wash it in 0.1× TBE buffer for 5 min with agitation.
8. Cross-link the RNA to the membrane using UV exposure of 4 mJ per second for 30 s.

3.4 Nonradioactive Detection of Small RNAs: Probe Generation

Isotopic detection of RNAs via northern blotting and hybridization incurs high costs and safety concerns. For this protocol we use a nonradioactive detection using digoxigenin-labelled RNA probes. Two different types of probes are applied depending on the nature of the endo-siRNAs to be detected. If multiple different endo-siRNAs are produced from the same precursor RNA we use a probe encompassing the entire region of complementarity between sense and antisense (*see* **Note 8**). If only a single sequence small RNA is to be detected we use a PCR generated poly-probe.

1. For the purposes of this protocol probe generation has been efficiently achieved using Ambion's mMESSAGE mMACHINE transcription kit using the suppliers instructions with digoxgenin-labelled nucleotide mixture used in place of the supplied nucleotides. Linearized plasmids containing full-length sense (2.4 kb) and antisense (1.3 kb) cDNA transcripts were used as templates for the labelling kit (*see* **Note 9**).
2. The poly-probe consists of a long oligonucleotide that contains 5 copies of the relevant small RNA sequence with short intervening linkers and a T7 and a SP6 adaptor on either end (Fig. 1).
3. The commercially synthesized poly-probe oligonucleotide is diluted and amplified by PCR using SP6 (5′-ATTTAGGTG ACACTATAGAA-3′) and T7 primers (5′-TAATACGACTCA CTATAGGG-3′).
4. The resulting fragment of ~200 bp must be purified because dNTPs interfere with the in vitro transcription (*see* **Note 10**). We use a PCR fragment purification kit and elute the DNA with water.
5. The DNA can be directly used for in vitro transcription according to the specific supplier's instructions (*see* **Note 11**). We aliquot the reaction and store the labelled probe at −80 °C. Use about 1 μL of the probe per 5 mL of hybridization solution.

3.5 Nonradioactive Detection of Small RNAs: Hybridization

1. Pre-hybridize the cross-linked membrane at 40 °C for >60 min in DIG Easy Hyb buffer.
2. Labelled probes require denaturation by heating at 75 °C for 10 min and then storing on ice.
3. Add the denatured probe to the pre-hybridization buffer and incubate with agitation overnight (*see* **Note 12**).
4. Post hybridization the membrane should be washed at low stringency: 0.1× SSC, 0.1 % SDS at room temperature (*see* **Note 13**).
5. For the chemiluminescent detection of bound DIG-probes, anti-DIG fab fragment linked to alkaline phosphatase enzymes were used in conjunction with ECL reagent according to the manufacturer's instructions.

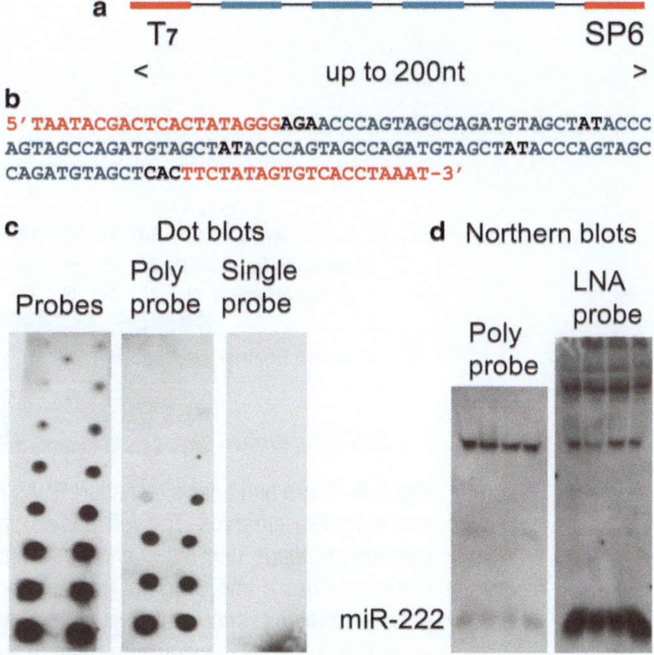

Fig. 1 Detection of miR-222 using a polyprobe. (**a**) Schematic representation of the polyprobe principle. The length of 200 bases results from limitations of commercial oligonucleotide synthesis. (**b**) Sequence of the miR-222 polyprobe; T7 and SP6 adaptors in *red*, miR-222 in *blue* and spacers in *black*. (**c**) Dot blot analysis of single and poly probe. The *left panel* shows dilutions of the single probe and polyprobe spotted on a nylon membrane. The *middle panel* shows T7 and SP6 RNA polymerase generated probes detecting decreasing amounts of target RNA. The *right blot* shows the same RNA spots but hybridized with the single probe. (**d**) Northern blots of human embryonic kidney cell, the four lanes contain different clones of the same cell line. The *left blot* was probed with the polyprobe against miR-222, the *right panel* shows to positive control using an locked nucleic acid (LNA) probe

4 Discussion

We used this protocol to assay for small RNAs associated with the conserved sodium phosphate co-transporter *Slc34a1* and its associated endogenous antisense transcript in microinjected *Xenopus* oocytes and in mouse tissue.

Small RNAs were extracted using TRIzol reagent incorporating some key modifications to the manufacturer's instructions. Total RNA samples were then separated by non-denaturing PAGE, blotted onto positively charged nylon membranes, hybridized, and detected using DIG-labelled RNA probes.

Small RNA signals were detected in mouse kidney (the physiological site for the Na–phosphate co-transporter) using the

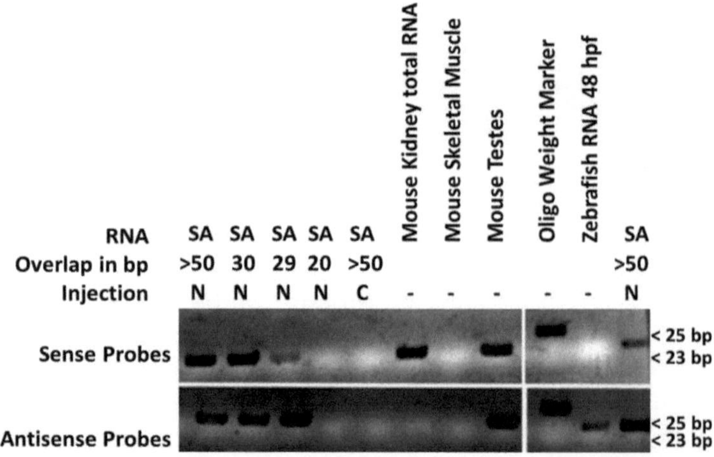

Fig. 2 Northern blot analysis of short RNAs from injected oocytes, mouse tissues, and zebrafish embryos. The *left panels* (*lanes 1–5*) show extracted RNA from injected *Xenopus* oocytes. *Lanes 1–3* indicate sense–antisense processing, whereas *lanes 4* and *5* do not contain short RNAs either because the overlap between injected sense–antisense transcripts is too short (*lane 4*) or the samples were injected into the cytoplasm (*lane 5*). *Lanes 6–8* show samples from mouse and *lanes 9–11* display control samples from zebrafish. *Lanes 6–8* represent tissues that express the sense encoded protein (kidney), do not express the transporter (skeletal muscle) or expresses both sense and antisense RNAs, but the presence of the transporter is unclear (testis). *Lane 10* represents short RNAs isolated from 48 h zebrafish embryos (19) as a positive control. *Lane 11* shows another control, i.e., a sample from *Xenopus* oocytes injected into the nucleus with sense and antisense RNA. "Sense"- and "antisense"-probes mean that the short RNAs detected with the sense probe will be complementary to the antisense transcript and vice versa. *SA* sense–antisense, *N* nuclear injection, *C* cytoplasmic injection, *hpf* hours post fertilization

sense-targeted probe; and in testis using both the sense- and antisense-targeted probes. Neither the sense nor antisense-targeted probes gave a signal in skeletal muscle. As controls, samples from zebrafish embryos were included (Fig. 2, right panel, lanes 9–11). These results concur with published data from zebrafish *Slc34a* wherein accumulation of strand-specific endo-siRNAs have been reported [4, 6].

5 Notes

1. For injected *Xenopus* oocytes 100 μL per single oocyte was adopted; for mouse tissues of <100 mg 1 mL of TRIzol was required. There are alternative methods to homogenize the tissue such as a small polytron mixer or a porcelain mortar and pestle to powder the N_2-frozen tissue; however,

cross-contamination may be an issue if multiple samples are being processed.

2. Incubation times should be adhered to so as to fully disrupt all protein-RNA associations. Due to the size difference of small RNAs and associated proteins, the incubation times are increased beyond the supplier's instructions to maximize disruption.

3. Efforts for further optimization of the protocol for a particular sample type would be best directed at the TRIzol–sample ratio and the centrifugation conditions.

4. The pre-running of the gel in the 1× TBE running buffer gives significantly better results in subsequent operations, e.g., northern blotting and detection.

5. The average MW of a small RNA will be in the range of 12,000–19,000 g/mol and is therefore equivalent to a 120 amino acid peptide. This allows the selection of an appropriate acrylamide percentage to be selected based upon the gel dimensions.

6. The inclusion of an RNA marker preparation can help in the orientation of the gel and for fragment size estimation purposes.

7. We use the membranes from Roche to be in line with the downstream DIG detection procedure which is a trademark of Roche. However, other products may also produce good results. The majority of membranes currently available are composed of a porous nylon matrix that is capable of maintaining a positive charge over a wide pH range (for example for BrightStar-Plus membranes the positive charge is stable in pH 3–10). This allows relatively strong ionic hybridizations to be achieved.

8. Overhangs on both ends increase signal intensity but may also cause higher background. For the ease of the experiment we just used the cloned full-length cDNAs for probe generation.

9. The mMESSAGE mMACHINE gives excellent result but is very pricy, in addition, it caps the RNA which is unnecessary for hybridization probes. It may be much cheaper to buy the RNA polymerases which come with the appropriate buffers and the DIG-labelled oligonucleotides.

10. Omission of this step results in no RNA being synthesized at all.

11. DIG-labelled nucleotides may not be as efficiently incorporated as unlabelled NTPs. In order to monitor efficiency we run a reaction without DIG-label in parallel. An aliquot of both reactions is then run on a normal 2.5 % agarose gel. The samples are denatured by heat and in RNA loading buffer,

but the gel is non-denaturing. The obtained resolution is sufficient to distinguish between the faster migrating unlabelled RNA and the slower DIG-labelled RNA.

12. The appropriate hybridization temperature should be calculated based upon the GC content of the probe and then subtracting ~20 °C, but this should be optimized for individual probes. For our probes, hybridization temperatures ranged between 37 and 40 °C.

13. Optimization of this step is key to the detection of small RNAs. Again, for our purposes washing in 0.1× SSC, 0.1 % SDS at temperatures ranging from RT to 40 °C gave positive results.

References

1. Katayama S, Tomaru Y, Kasukawa T, Waki K, Nakanishi M, Nakamura M, Nishida H, Yap CC, Suzuki M, Kawai J, Suzuki H, Carninci P, Hayashizaki Y, Wells C, Frith M, Ravasi T, Pang KC, Hallinan J, Mattick J, Hume DA, Lipovich L, Batalov S, Engstrom PG, Mizuno Y, Faghihi MA, Sandelin A, Chalk AM, Mottagui-Tabar S, Liang Z, Lenhard B, Wahlestedt C (2005) Antisense transcription in the mammalian transcriptome. Science 309:1564–1566
2. Baulcombe D (2004) RNA silencing in plants. Nature 431:356–363
3. Kim VN, Han J, Siomi MC (2009) Biogenesis of small RNAs in animals. Nat Rev Mol Cell Biol 10:126–139
4. Carlile M, Swan D, Jackson K, Preston-Fayers K, Ballester B, Flicek P, Werner A (2009) Strand selective generation of endo-siRNAs from the Na/phosphate transporter gene Slc34a1 in murine tissues. Nucleic Acids Res 37:2274–2282
5. Chomczynski P, Sacchi N (1987) Single-step method of RNA isolation by acid guanidinium thiocyanate-phenol-chloroform extraction. Anal Biochem 162:156–159
6. Carlile M, Nalbant P, Preston-Fayers K, McHaffie GS, Werner A (2008) Processing of naturally occurring sense/antisense transcripts of the vertebrate Slc34a gene into short RNAs. Physiol Genomics 34:95–100

Chapter 9

p19-Mediated Enrichment and Detection of siRNAs

Jingmin Jin, Larry A. McReynolds, and Monika Gullerova

Abstract

p19 is an RNA binding protein originally isolated from the Carnation Italian ring-spot virus (CIRV). It has been shown that p19 is a plant RNA-silencing suppressor that binds small interfering RNA (siRNA) with high affinity. A bifunctional p19 fusion protein, with an N-terminal maltose binding protein (MBP) and a C-terminal chitin binding domain (CBD) allows protein purification and binding of p19 to chitin magnetic beads via the chitin binding domain. The fusion p19 protein recognizes and binds double-stranded RNAs (dsRNA) in the size range of 20-23 nucleotides, but does not bind single strand RNA (ssRNA) or dsDNA. Furthermore, p19 can also bind mRNA, if there is a 19 bp blunt RNA duplex at the exact end of the RNA. Binding specificity of the p19 fusion protein for small dsRNA allows for detection of siRNAs derived either from exogenous or endogenous long dsRNA or microRNAs when hybridized to a complementary RNA. Here we describe a robust method using p19 and radioactive RNA probes to detect siRNAs in the sub-femtomole range and in the presence of a million-fold excess of total RNA. Unlike most nucleic acid detection methods, p19 selects for RNA hybrids of correct length and structure. This chapter describes the potential of p19 fusion protein to detect miRNAs, isolate exogenous or endogenous siRNAs, and purify longer RNAs that contain a 19-bp terminal RNA duplex.

Key words Double-stranded RNA, Small interfering RNA, p19 protein, Radioactively labeled RNA probe, p19 protein pull down, Northern blot

1 Introduction

RNA interference (RNAi) and microRNA (miRNA)-mediated gene silencing pathways are key regulatory mechanisms controlling gene expression in eukaryotes [1, 2]. Both pathways depend on 20- to 25-nt small interfering RNAs (siRNAs) that derive either from longer double-stranded RNA (siRNAs) or hairpin precursors (miRNAs) that are processed by the RNase III-type endonucleases Drosha and Dicer [3]. Together with other proteins, siRNAs form part of effector complexes that consequently trigger translational repression and/or degradation of target RNAs, resulting in changes in gene expression at the posttranscriptional (PTGS) [4] or transcriptional level (TGS) [5].

Andreas Werner (ed.), *Animal Endo-siRNAs: Methods and Protocols*, Methods in Molecular Biology, vol. 1173, DOI 10.1007/978-1-4939-0931-5_9, © Springer Science+Business Media New York 2014

Currently existing methods for siRNA detection, which include microarrays [6–8], single-molecule detection with lasers [9], or microsphere detection of miRNAs [10], often require expensive equipment. Other methods that use multiple enzymatic steps including RNA ligation, amplification, and gel electrophoresis [11–14] are difficult to quantify. In this chapter, we describe a novel method for siRNA detection that utilizes the siRNA binding properties of the p19 protein from the Carnation Italian ringspot virus (CIRV), a plant tombusvirus. The RNA-binding features of the protein were discovered as a result of CIRV's ability to block RNAi in plants [15–17]. The dimeric p19 binds dsRNA in a size-dependent, sequence-independent manner [18, 19].

Hybridization of a siRNA-specific probe to a single-stranded target miRNA or siRNA creates dsRNA that tightly binds to p19 protein which, for experimental purposes, has been modified: The amino terminus of p19 is fused to maltose binding protein (MBP) to aid purification [20] and the carboxy terminus is fused to chitin binding domain (CBD) for tight binding to chitin magnetic beads [21] (Fig. 1). The recombinant p19 fusion protein retains the ability to selectively bind siRNAs and shows the greatest affinities for 19- and 20-nucleotide dsRNAs with blunt ends, as well as for 21-nucleotide siRNAs with 5′ phosphates and 3′ 2-nucleotide extensions. This novel siRNA detection method is linear over a three-log concentration range and has a sensitivity of 0.1 fmol (Fig. 2). It is also significantly faster and more sensitive than northern blot analysis.

2 Materials

1. New England Biolab provides a kit that contains purified p19 protein as well as paramagnetic BSA-treated chitin beads.
2. 1× p19 siRNA binding buffer containing 20 mM Tris–HCl, 100 mM NaCl, 1 mM EDTA (ethylenediaminetetraacetic acid), 1 mM TCEP [tris(2-carboxyethyl)phosphine hydrochloride)], 0.02 % Tween-20, pH 7.0 at 25 °C.
3. p19 siRNA 1× washing buffer containing 20 mM Tris–HCl, 100 mM NaCl, 1 mM EDTA, 1 mM TCEP, pH 7.0 at 25 °C, supplemented with 100 μg/mL of BSA (bovine serum albumin).
4. p19 siRNA 1× elution buffer containing 20 mM Tris–HCl, 100 mM NaCl, 1 mM EDTA, 1 mM TCEP, 0.5 % SDS, pH 7.0 at 25 °C.
5. Magnetic rack to attract and collect magnetic beads at the wall or bottom of tubes.
6. T4 Polynucleotide Kinase for the labeling of the probe RNA oligonucleotide. This enzyme comes with the 10× buffer (New England Biolabs).

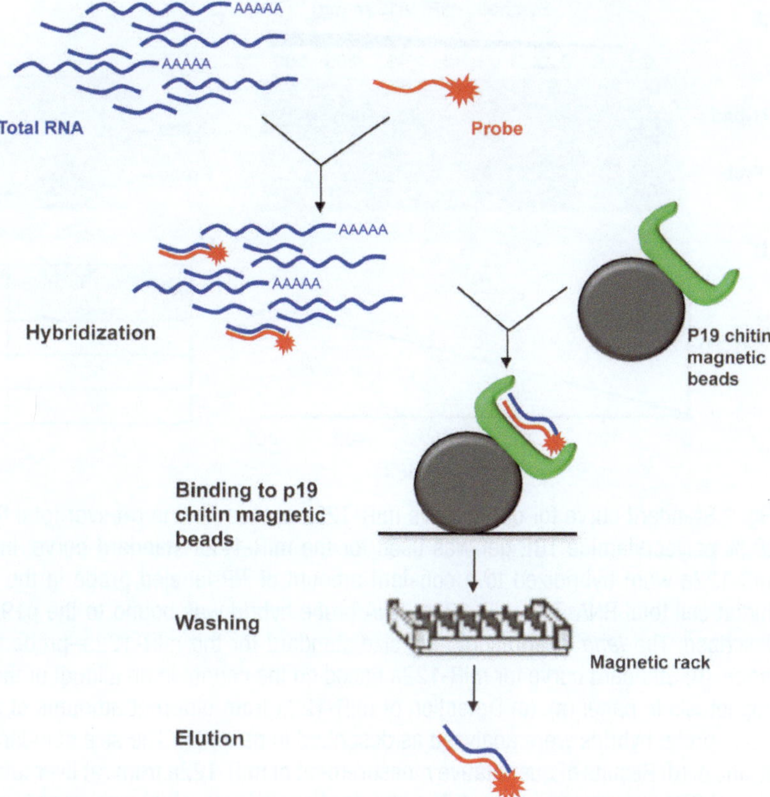

Fig. 1 miRNA detection using p19 beads. The radioactively labeled miRNA probe (*red*) is hybridized with a total RNA extract containing target miRNAs and bound to the p19 chitin magnetic beads. The p19 fusion protein (*green*) is linked to the beads through the chitin-binding domain. The unbound probe is removed by washing and the miRNA–probe duplex can be eluted and quantitatively measured

7. γ-^{32}P-ATP with a specific activity of 6,000 Ci/mM. ^{32}P is a beta emitter and working with the substance requires adequate protection. Adhere to the guidelines of handling radioactive substances.
8. CentriSep columns (Princeton Separation) to purify the labeled oligonucleotides form unincorporated nucleotides and salts.
9. Denaturing polyacrylamide gel electrophoresis and northern blotting are not the focus of this article, but are discussed in detail in subheading 3.4. We used the following materials: 20 % acrylamide non-denaturing TBE gel or 15 % TBE urea gel, 2× TBE-urea sample buffer, and yeast RNA as a carrier (Invitrogen Life Science Technologies). Blotting was performed using a Semi-Dry Electro blotter (Continental Lab Products, San Diego, CA, USA) in 1× TBE buffer, Gene screen Plus membrane (Perkin Elmer Life and Analytical Sciences) and a Spectrolinker XL-1000 UV crosslinker

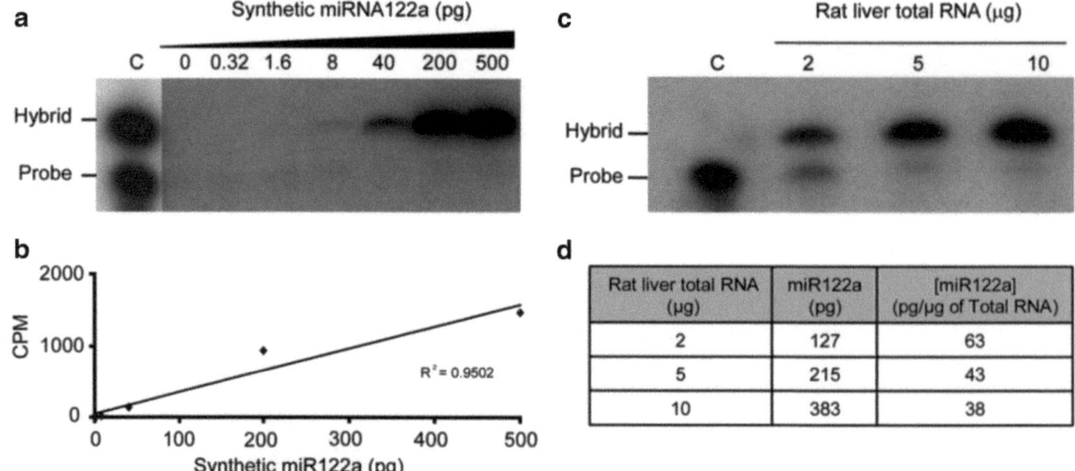

Rat liver total RNA (μg)	miR122a (pg)	[miR122a] (pg/μg of Total RNA)
2	127	63
5	215	43
10	383	38

Fig. 2 Standard curve for quantitative miR-122a assessment in rat liver total RNA. (**a**) An autoradiograph of a 20 % polyacrylamide TBE gel was used for the miR-122a standard curve. Increasing amounts of synthetic miR-122a were hybridized to a constant amount of ^{32}P-labeled probe in the presence of a large excess of Jurkat cell total RNA. The miR-122a–RNA probe hybrid was bound to the p19 beads, washed, and eluted as described. The *lane C* represents the size standard for the miR-122a–probe hybrid and the single stranded probe. (**b**) Standard curve for miR-122a based on the counts in an aliquot of the eluted RNA used for gel electrophoresis in panel (**a**). (**c**) Detection of miR-122a from different amounts of rat liver total RNA. Eluted miR-122a–probe hybrids were analyzed as described in panel (**a**). The size standard for RNA probe is represented in lane **c**. (**d**) Results of quantitative measurement of miR-122a from rat liver total RNA. Three different amounts of total RNA samples were used. The standard curve in panel (**b**) was used to convert cpm to pg of miR-122a

(Spectronics Corporation, Westbury, NY, USA). (Pre) Hybridization was done in Ultrahyb-Oligo hybridization buffer (Invitrogen Life Science Technologies). After stringency washes the blot was exposed to Hyperfilm MP (GE Healthcare).

10. Murine RNase inhibitor (New England Biolabs).
11. 4 M KCl: 29.824 g of KCl in a final volume of 100 mL of water.
12. AMV First Strand cDNA Synthesis Kit (New England Biolabs).

3 Methods

3.1 Protocol for Double-Strand siRNA Detection

1. Binding of siRNA to p19 siRNA Binding Protein: Add 50 U of p19 protein to the RNA extract (>1 μg) containing siRNAs or double-stranded miRNAs in 1× p19 siRNA binding buffer.
2. Incubate the mix for 1 h at room temperature with shaking to form the siRNA–p19 complex.
3. Binding of siRNA–p19 complex to chitin magnetic beads: Resuspend the BSA treated chitin magnetic beads well and aliquot 20 μL of the suspension into a sterile microfuge tube.

4. Pull the beads to the side of the tube with a magnetic rack and remove the supernatant.
5. Add the siRNA–p19 complex to the beads and mix by shaking or laying the tube on a magnetic stir plate for 1 h at room temperature.
6. To remove unbound RNA place the tube into the magnetic rack and carefully remove the supernatant containing the unbound RNA.
7. Wash the beads with 500 μL of p19 siRNA 1× washing buffer. Vortex and collect the beads using the magnetic stand. Discard the supernatant.
8. Repeat the washing step three times, and after third wash, remove as much buffer as possible without disturbing the pellet.
9. *Elute the siRNA by adding* 30–40 μL of p19 siRNA 1× elution buffer to the bead pellet.
10. Incubate at 37 °C for 10 min.
11. Mix on a magnetic stirring plate for 10 min at room temperature. Collect the beads using the magnetic rack and remove the siRNA-containing supernatant. The elution step can be repeated and eluted samples combined to increase the yield.
12. The isolated siRNAs are now ready for further analysis and can be visualized by northern blotting.

3.2 Protocol for Single-Stranded miRNA or siRNA Detection

miRNA or siRNA probe design: Most miRNAs and siRNAs are 19–25 bases in length and single stranded. To detect siRNAs/miRNAs using p19 beads they must be hybridized to the complementary RNA strand. The optimum size of double-stranded RNA is 19 nucleotides with perfect complementarity between probe and target siRNA/miRNA. The RNA probe ideally forms a blunt ended duplex with one end of the siRNA/miRNA. The RNA oligonucleotide probe should not have a 5′ phosphate so it can be labeled (*see* following section).

1. Labeling of the RNA probe: Combine the following reaction in a sterile microcentrifuge tube:

   20–100 ng of the probe RNA oligonucleotide with a 5′ hydroxyl group.

   2 μL of T4 Polynucleotide Kinase Buffer (10×).

   20–100 μCi of γ-^{32}P-ATP 2 μL T4 Polynucleotide Kinase (10 U/μL).

   Add water to 20 μL.
2. Incubate the reaction at 37 °C for 60 min followed by 20 min at 65 °C to inactivate the enzyme. Use CentriSep column (Princeton Separation) to remove the unincorporated isotope,

follow the supplier's protocol. Measure the specific activity of the purified oligonucleotide using a scintillation counter.

3. To allow for quantitative detection of siRNA/miRNAs, a standard curve is made using decreasing amounts of a synthetic siRNA (from 300 to 1 pg) which are hybridized to a constant amount of the ^{32}P-labeled probe (~500 pg). The hybridization should contain several micrograms of total RNA from yeast or bacteria, which does not contain the siRNA of interest to mimic siRNA/miRNA detection conditions in the unknown samples. The counts bound and eluted from the p19 beads are analyzed using the standard curve and linear regression.
4. Hybridization of labeled RNA probes to siRNA/miRNA. Typically, 1–0.5 ng of ^{32}P-labeled probe is hybridized to a sample containing total RNA (from 1 to 10 μg) by incubating at 55–65 °C for 2 h in 10 μL containing 1× p19 binding buffer. The hybridization temperature depends on the melting temperature of the probe–siRNA hybrid. Hybridization allows the probe to form duplex with the endogenous target siRNA/miRNA which binds to p19.
5. During the hybridization step prepare the complex of p19 with BSA treated chitin magnetic beads. Resuspend the BSA treated chitin magnetic beads by briefly vortexing and transfer 10 μL of the suspension into a 1.5 mL microcentrifuge tube.
6. Add 3 μL of p19 protein to the 10 μL beads aliquot and incubate it on benchtop shaker for 10–20 min at room temperature.
7. Prepare the RNA binding buffer by adding the following components into a 1.5 mL microcentrifuge tube: 6 μL of RNase-free water, 1 μL of 10× p19 binding buffer, 2 μL of 10× BSA (dilute from the 100× stock), and 1 μL of RNase inhibitor.
8. Mix briefly and add 10 μL RNA binding buffer to the RNA hybridization reaction to bring the final volume to 20 μL.
9. Binding of the hybrid RNA duplex to the p19 beads: Remove the supernatant from the p19 chitin beads (**step 5**) using the magnetic separation rack. Add the entire 20 μL of hybridization mix to the beads and incubate the reaction on a shaker for 1.5 h at room temperature.
10. An excess of probe and unbound RNA are removed by washing the p19 beads with pre-warmed BSA-wash buffer (20 mM Tris–HCl, 100 mM NaCl, 1 mM EDTA, 1 mM TCEP, 1 mg/mL BSA, pH 7.0 at 25 °C). Remove the supernatant from the binding reaction using the magnetic separation rack. Wash the beads in 500 μL of 1× BSA-wash buffer and incubate on a benchtop shaker for 5 min at room temperature. Repeat the wash five times. To minimize any loss of the beads during the wash steps, allow the beads to settle to the bottom of the tube

before using the magnetic rack. Beads are drawn to the side of the tube using the magnetic rack and the supernatant is carefully removed.

11. After the last wash, remove as much of the supernatant as possible without touching the beads pellet.
12. Add 40 μL of pre-warmed 1× p19 elution buffer to the beads and incubate for 10 min at room temperature with shaking followed by a 10 min incubation at 37 °C.
13. Collect the beads with the magnetic rack and carefully remove the supernatant containing the RNA hybrids from the microfuge tube. 5–10 μL of the eluate is usually sufficient for analysis of the sample on a liquid scintillation counter or PAGE.

3.3 Detection of miR-122a with p19 Beads

To validate the sensitivity and specificity of miRNA detection using p19 magnetic beads we determined the amount of miR-122a in rat liver RNA (Molecular Research Center, Cincinnati, OH, USA).

1. A RNA probe complementary to miR-122a, 5′-AACACCA UUGUCACACUCCAUA-3′, was end-labeled with 6,000 Ci/mM [γ-^{32}P]-ATP using T4 polynucleotide kinase as described in Subheading 3.2.
2. Different amounts of total rat liver RNA (5 and 10 μg) were hybridized with 1 ng (500,000 cpm) of ^{32}P-labeled miR-122a probe at 75° for 5 min and 52° for 7 h in a programmed PCR machine in 1× p19 binding buffer in a total volume of 10 μL.
3. The hybridization mixture was then bound to 10 μL p19 beads, washed, and eluted as described above.
4. A 1/50 aliquot of the eluted miR-122a:RNA probe duplex, which represents 100 and 200 ng of rat liver RNA, was analyzed by electrophoresis on a 20 % TBE gel. The gel was dried and exposed to an Imaging Screen K (Bio-Rad).
5. The amount of radioactivity in the hybrid was determined by a scan of the screen using a Typhoon 9400 scanner with Image Quant TL software (GE Healthcare). A standard curve for miR-122a detection was generated by hybridizing a 1-ng aliquot of the same radioactive miR-122a probe with different amounts (20–0.1 fmol) of synthetic miR-122a (5′-pUG-GAGUGUGACAAUGGUGUUUG) in the presence of 10 μg yeast total RNA (Invitrogen). The same p19 binding protocol was used for the standard curve and rat liver RNA samples. All assays were done in triplicate (Fig. 2).

3.4 The Sensitivity of p19 Compared to a Northern Blot for Detection of miR-153

Some miRNAs are present in low amounts and therefore very difficult to detect. An example is miRNA-153 from the parasitic nematode, *Brugia malayi*. A comparison was made between the p19 beads method and a standard northern blot analysis to detect this miRNA. The same radioactive RNA probe and the same time of exposure of the film were used in both assays.

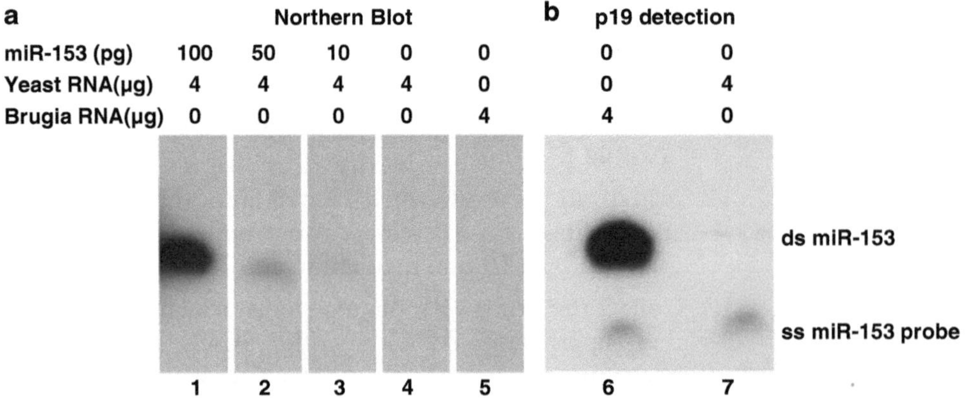

Fig. 3 Comparison of p19 enrichment and northern blots for miR-153 detection. A synthetic RNA oligonucleotide complementary to miR-153 was radiolabeled with ^{32}P and was used for both northern blot probe and miR-153 complement in p19 purification. (**a**) Northern blot detection of miR-153. *Lanes 1–3* contain 4 μg of yeast total RNA mixed with 100, 50, or 10 pg of synthetic miR-153 RNA oligonucleotide. *Lane 4* contains 4 μg of yeast total RNA without miRNA (negative control). *Lane 5* contains 4 μg of *B. malayi* total RNA. (**b**) Detection of miR-153 using p19 beads. The ^{32}P-labeled miR-153 RNA probe was hybridized to either 4 μg of *B. malayi* total RNA or to 4 μg of yeast total RNA as a negative control. After binding to p19 beads, the samples were washed, eluted analyzed on a 20 % TBE non-denaturing gel. *Lane 6* contains *B. malayi* RNA and *lane 7* yeast RNA

Northern blot detection of miR-153:

1. RNA samples were prepared in 2× TBE-urea sample buffers and loaded on a 15 % TBE-urea gel. Each gel lane contained either 4 μg yeast RNA or 4 μg RNA isolated from the microfilariae of the parasitic nematode *Brugia malayi* (TRS Labs, Athens, GA, USA). Positive controls contained 4 μg yeast RNA and 100, 50, or 10 pg miR-153 RNA primers.
2. Following electrophoresis, the RNA was electroblotted onto Genescreen Plus membrane using a Semi-Dry Electro blotter in 1× TBE, as described by the manufacturer, and then UV cross-linked to the membrane at 120 mJ/cm^2.
3. After pre-hybridization in Ultrahyb-Oligo hybridization buffer, the membrane was hybridized overnight at 37 °C in the same buffer containing ^{32}P-labeled oligonucleotide probes (2×10^6 cpm/mL hybridization solution). The 19-nucleotide long probe, 5′-CACUUUUGUGACUAUGCAA (complementary to miR-153) was labeled with [γ-^{32}P]-ATP and T4 polynucleotide kinase as described in Subheading 3.2. The blot was washed twice at 42 °C for 30 min each in 2× SSC, 0.1 % SDS. The blot was exposed to Hyperfilm MP (GE Healthcare) for 3 days at −70 °C using an intensifying screen (Fig. 3).

p19 detection of miR-153

1. 1 ng of the ^{32}P-labeled northern blot RNA probe was hybridized to either 4 μg *B. malayi* RNA or 4 μg of yeast RNA for 2 h at 65 °C.
2. After hybridization 10 μL of p19 beads were added to the hybridization reaction to form the RNA–p19–beads complex by shaking at room temperature for 2 h. The non-hybridized probe was removed by washing the beads as described in Subheading 3.1.
3. The duplex RNA was eluted from the beads with SDS and separated by electrophoresis using a 20 % acrylamide non-denaturing TBE gel. The gel was dried and exposed to Hyperfilm MP for 3 days at −70 °C using an intensifying screen.

The northern blot (Fig. 3a) shows that in the positive control 50 pg of a synthetic miR-153 can be detected, however, no signal was observed in the *B. malayi* RNA lane. In the p19 assay a very strong signal was detected in the *B. malayi* sample and none in the negative control of the yeast RNA (Fig. 3b). This suggests that p19 assay is at least 50 times more sensitive than the northern blot.

3.5 Detection of SL1 mRNAs and SL2 mRNAs from C. elegans Total RNA

p19 binding to RNA is not restricted to short dsRNAs: A RNA oligonucleotide can form a blunt ended duplex exactly at the 5′ end of a long mRNA to produce a p19 substrate. The ability of p19 to isolate mRNA by this method was tested with RNA complementary to the spliced leader in *C. elelgans.* About 70 % of *C. elegans* pre-mRNAs are trans-spliced to one of two 22 nucleotide spliced leaders (SL1 and SL2). Among these trans-spliced mRNAs, most are SL1 trans-spliced and only less than 20 % are SL2 trans-spliced. Trans-splicing occurs throughout the nematode phylum as well as in some protists (e.g., trypanosomes) and many other animals including flatworms, hydra, and primitive chordates [22].

The p19 magnetic beads were used to specifically pull down either SL1 or SL2 trans-spliced mRNAs from a total RNA prep. The specificity of the pull down was tested by reverse-transcription PCR.

Annealing of synthetic 19 nucleotides long SL1 RNA probe

1. Mix 10 μg of *C. elegans* total RNA with 100 ng of the SL1 probe (30 ng for SL2 probe) in 1× p19 binding buffer in a total volume of 10 μL. Heat the mix at 95 °C for 5 min in a heating block, then turn off the block and let it cool down slowly to room temperature.
2. Quickly spin the annealing mix and add 2 μL of 10× p19 binding buffer, 2 μL of 10× BSA, and 2 μL of murine RNase inhibitor and bring the total volume to 30 μL with water. Take 20 μL of p19 beads, pull them to the side of the tube with a magnetic

rack and then remove the supernatant. Resuspend the p19 beads with 30 μL of the annealed RNA. Shake the RNA–beads mix in a benchtop shaker for 2 h at room temperature.

3. Wash the p19 beads as described in Subheading 3.1. To elute the trans-spliced mRNAs from the beads, mix 50 μL of 1× elution buffer with the beads pellet and shake for 30 min at 37 °C. Collect the supernatant using the magnetic rack.
4. To remove the SDS from the eluate, add in 4 μL of 4 M KCl and mix well. The eluate becomes cloudy as a result of the SDS precipitate. Incubate the eluate at 4 °C for 2 h and centrifuge for 30 min at maximum speed in a microcentrifuge. Carefully remove the supernatant without touching the white SDS pellet, and recentrifuge if necessary.
5. To test the specificity of the p19 pull down RT-PCR is performed to selectively amplify SL1 or SL2 trans-spliced RNA.
6. First strand cDNA is synthesized using the Protoscript kit according to the supplier's protocol and 12 μL of SL1 and SL2 enriched RNA. Mix the RNA with 4 μL of the random primer and heat it to 70 °C, for 5 min. Return the tube on ice and add 20 μL of AMV mix, 4 μL of enzyme mix and incubate at 25 °C for 5 min and then at 42 °C incubation for 1 h. The reaction is stopped by heat inactivation at 80 °C for 5 min. For PCR, we dilute the cDNA by adding 50 μL of water.
7. For this experiment, two *C. elegans* SL1 trans-spliced genes, Kpc-1 and Pgd-4, and two SL2 trans-spliced genes, Tra-2 and Kin10, were chosen for amplification from the two SL enriched pools. A gene specific reverse primer, located 200–400 bp downstream from the splice leader sequence for each of the four genes, was combined with either SL1 or SL2 primer to PCR amplify a fragment of the trans-spliced genes.
8. The PCR products were analyzed on a 10 % TBE polyacrylamide gel. As shown, in Fig. 4b only SL1 trans-spliced genes, Kpc-1 and Pgd-4, can be amplified from the SL1 pull down sample, while SL2 trans-spliced genes, Tra-2 and Kin-10, can only be amplified from the SL2 pull down sample. This demonstrates the ability of the p19 beads to isolate longer RNAs when they contain a 5′ dsRNA duplex. This approach can therefore be used to specifically enrich RNAs with a known 5′ sequence.

3.6 Protocol for Enrichment of siRNA with p19 Beads

1. Since siRNAs are often present at very low amounts we performed a reconstitution experiment, where a synthetic siRNA primer was added to a large excess of total RNA. The siRNA was then isolated using p19 magnetic beads. 20 ng of siRNA were mixed with 55 μg of total rat liver RNA in a total volume

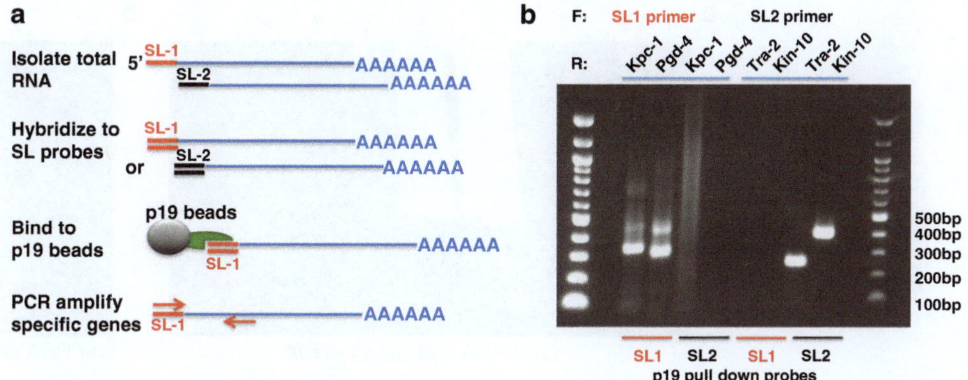

Fig. 4 Selective pull downs of trans-spliced mRNAs by p19 beads. (**a**) The diagram shows selective binding of p19 beads to RNA with a 5′ RNA duplex. Only trans-spliced *C. elegans* mRNAs are shown. Splice leader 1, SL1 (*red*), and splice leader 2, SL2 (*black*), are at the 5′ end of the mRNAs. The double lined regions (*red* or *black*) show SL1 and SL2 probes hybridized to the splice leaders. The p19 protein (*green*) attached to magnetic beads (*gray*) binds the dsRNA at the 5′ end of the mRNA. The eluted mRNA is then amplified with two PCR primers, indicated by *two red arrows*. One primer is specific for SL1 and the reverse primer is a gene specific. (**b**) The PCR products for the four genes that are pulled down with either SL1 or SL2 are resolved on a TBE acrylamide gel. F indicates SL1 or SL2 PCR forward primers and R represents the reverse primers that are gene specific. SL1 (*red*) or SL2 (*black*) at the *bottom* of the figure indicate which complementary RNA probes were used for the specific pull downs

of 40 μL. A negative control contained 55 μg of liver RNA and no siRNA. Half of each sample (20 μL) was for p19 mediated siRNA enrichment as described in Subheading 3.1.

2. p19 beads were washed and the siRNA was eluted with SDS and the entire samples were run on a 20 % TBE gel. 20 μL of liver RNA, an aliquot of the initial siRNA as well as first and last washes were analyzed in parallel. The results show that a large excess of total RNA is removed during the washing steps (Fig. 5, lane 5 compared to lane 8). Moreover, the eluted RNA contains only siRNAs (Fig. 5 lane 8), no siRNA signal is seen in samples that just contain liver RNA (Fig. 5b). This demonstrates that p19 beads can enrich siRNAs at least 1,000-fold with good yields.

Acknowledgement

This work was supported by a MRC Career Development Award to M.G. New England Biolabs supported the research of J.J. and L.M.

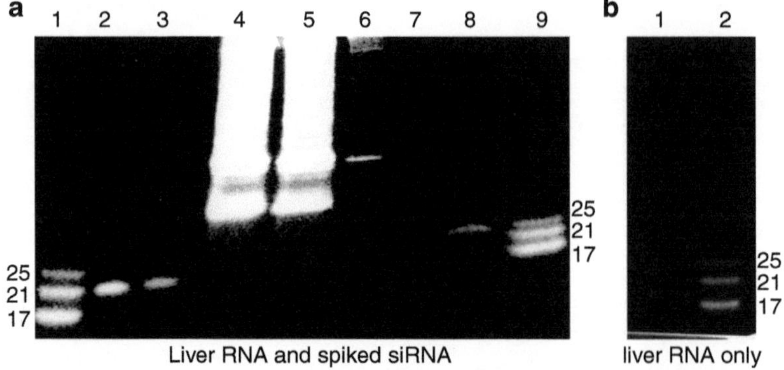

Fig. 5 Efficient enrichment of siRNA from rat liver total RNA. (**a**) *Lanes 1* and *9* are dsRNA size markers. *Lanes 2* and *3* contain 20 and 10 ng of siRNA as a quantitative control. *Lane 4*, 20 μL of liver RNA and siRNA mix (starting material). *Lane 5*, 20 μL of total RNA not bound to the beads (*1* supernatant). *Lane 6* and *7*, wash 1 and 6 of the p19 beads (20 of 500 μL wash buffer). *Lane 8*, 20 μL of the eluted siRNAs. (**b**) *Lane 1*, 20 μL of the eluate from the negative control (p19 beads mixed with 27.5 μg of rat liver RNA and without siRNA added control)

References

1. Chen K, Rajewsky N (2007) The evolution of gene regulation by transcription factors and microRNAs. Nat Rev 8:93–103
2. Bartel DP (2009) MicroRNAs: target recognition and regulatory functions. Cell 136: 215–233
3. Bernstein E, Caudy AA, Hammond SM, Hannon GJ (2001) Role for a bidentate ribonuclease in the initiation step of RNA interference. Nature 409:363–366
4. Fire A, Xu S, Montgomery MK, Kostas SA, Driver SE, Mello CC (1998) Potent and specific genetic interference by double-stranded RNA in Caenorhabditis elegans. Nature 391:806–811
5. Morris KV, Chan SW, Jacobsen SE, Looney DJ (2004) Small interfering RNA-induced transcriptional gene silencing in human cells. Science 305:1289–1292
6. Krichevsky AM, King KS, Donahue CP, Khrapko K, Kosik KS (2003) A microRNA array reveals extensive regulation of microRNAs during brain development. RNA 9: 1274–1281
7. Bentwich I, Avniel A, Karov Y, Aharonov R, Gilad S, Barad O, Barzilai A, Einat P, Einav U, Meiri E, Sharon E, Spector Y, Bentwich Z (2005) Identification of hundreds of conserved and nonconserved human microRNAs. Nat Genet 37:766–770
8. Shingara J, Keiger K, Shelton J, Laosinchai-Wolf W, Powers P, Conrad R, Brown D, Labourier E (2005) An optimized isolation and labeling platform for accurate microRNA expression profiling. RNA 11:1461–1470
9. Neely LA, Patel S, Garver J, Gallo M, Hackett M, McLaughlin S, Nadel M, Harris J, Gullans S, Rooke J (2006) A single-molecule method for the quantitation of microRNA gene expression. Nat Methods 3:41–46
10. Lu J, Getz G, Miska EA, Alvarez-Saavedra E, Lamb J, Peck D, Sweet-Cordero A, Ebert BL, Mak RH, Ferrando AA, Downing JR, Jacks T, Horvitz HR, Golub TR (2005) MicroRNA expression profiles classify human cancers. Nature 435:834–838
11. Schmittgen TD, Jiang J, Liu Q, Yang L (2004) A high-throughput method to monitor the expression of microRNA precursors. Nucleic Acids Res 32:e43
12. Chen C, Ridzon DA, Broomer AJ, Zhou Z, Lee DH, Nguyen JT, Barbisin M, Xu NL, Mahuvakar VR, Andersen MR, Lao KQ, Livak KJ, Guegler KJ (2005) Real-time quantification of microRNAs by stem-loop RT-PCR. Nucleic Acids Res 33:e179
13. Jiang J, Lee EJ, Gusev Y, Schmittgen TD (2005) Real-time expression profiling of microRNA precursors in human cancer cell lines. Nucleic Acids Res 33:5394–5403

14. Maroney PA, Chamnongpol S, Souret F, Nilsen TW (2007) A rapid, quantitative assay for direct detection of microRNAs and other small RNAs using splinted ligation. RNA 13:930–936
15. Silhavy D, Molnar A, Lucioli A, Szittya G, Hornyik C, Tavazza M, Burgyan J (2002) A viral protein suppresses RNA silencing and binds silencing-generated, 21- to 25-nucleotide double-stranded RNAs. EMBO J 21:3070–3080
16. Voinnet O, Pinto YM, Baulcombe DC (1999) Suppression of gene silencing: a general strategy used by diverse DNA and RNA viruses of plants. Proc Natl Acad Sci U S A 96:14147–14152
17. Lakatos L, Szittya G, Silhavy D, Burgyan J (2004) Molecular mechanism of RNA silencing suppression mediated by p19 protein of tombusviruses. EMBO J 23:876–884
18. Ye K, Malinina L, Patel DJ (2003) Recognition of small interfering RNA by a viral suppressor of RNA silencing. Nature 426:874–878
19. Vargason JM, Szittya G, Burgyan J, Hall TM (2003) Size selective recognition of siRNA by an RNA silencing suppressor. Cell 115:799–811
20. Calabrese JM, Sharp PA (2006) Characterization of the short RNAs bound by the P19 suppressor of RNA silencing in mouse embryonic stem cells. RNA 12:2092–2102
21. Chong S, Mersha FB, Comb DG, Scott ME, Landry D, Vence LM, Perler FB, Benner J, Kucera RB, Hirvonen CA, Pelletier JJ, Paulus H, Xu MQ (1997) Single-column purification of free recombinant proteins using a self-cleavable affinity tag derived from a protein splicing element. Gene 192:271–281
22. Allen MA, Hillier LW, Waterston RH, Blumenthal T (2011) A global analysis of C. elegans trans-splicing. Genome Res 21:255–264

Chapter 10

Detection of Small Noncoding RNAs by In Situ Hybridization Using Probes of 2′-O-Methyl RNA + LNA

Martin Jensen Søe, Martin Dufva, and Kim Holmstrøm

Abstract

In situ hybridization is a powerful method to provide information about contextual distribution and cellular origin of nucleic acids, e.g., in formalin-fixed paraffin-embedded (FFPE) samples of tissue. Particularly the recently discovered classes of noncoding RNA (ncRNA) including endo-siRNAs and microRNAs require such a technique to enable their study and visualization in natural contexts, and in the last decade, many advances have been made, increasing our ability to specifically detect small ncRNAs. One of the key developments has been the demonstration of the superiority of using locked nucleic acid (LNA)-modified DNA probes for the detection of ncRNA in tissue. Here, we describe an alternative in situ hybridization protocol employing oligonucleotide probes consisting of combinations of LNA and 2′-O-methyl RNAs that under optimized hybridization buffer conditions can provide a highly sensitive assay performance with only 1 h hybridization time.

Key words Noncoding RNA, microRNA, LNA, In situ hybridization, 2′-O-methyl RNA

1 Introduction

In the last decade, significant developments towards detecting short noncoding RNAs, e.g., microRNAs, by in situ hybridization (ISH) techniques have been achieved using locked nucleic acid (LNA)-modified DNA probes [1–9]. LNAs are RNA nucleotides containing a methylene bridge between the 2′ oxygen and 4′carbon, locking the ribose in a C3′endo conformation [10, 11]. Performing miRNA ISH using DNA probes with LNA substitutions (LNA + DNA) has been shown to stabilize the resulting duplexes compared to unmodified DNA probes and simultaneously improve discrimination between perfect matched and mismatched targets [1, 12]. More elaborate and laborious procedures have been published for the detection of low-copy number miRNAs including the introduction of an EDC fixation step to cross-link the 5′ end of miRNA to the tissue section [5], lengthy hybridizations (7–15 h) which use either low stringency washing (4 °C) with

Andreas Werner (ed.), *Animal Endo-siRNAs: Methods and Protocols*, Methods in Molecular Biology, vol. 1173,
DOI 10.1007/978-1-4939-0931-5_10,

probe concentrations of 50 nM, or, alternatively, high stringency washing (50 °C) with probe concentrations of 400–2,000 nM [4]. An extensively modified version of the protocol by Pena et al. [5] was recently published, which among other things eliminated the need for proteinase K treatment for tissue permeabilization, but still relied on lengthy overnight hybridizations (16 h) and the use of toxic denaturant formamide to achieve sufficient sensitivity [9].

Here, we describe an alternative in situ hybridization method where we use 2′-O-methyl RNA (2OMe) substituted LNA-containing oligonucleotide probes, that in some instances and under certain hybridization buffer conditions allow highly sensitive and fast (1 h) hybridization times. We have shown that the substitution of DNA with 2OMe nucleotides in LNA-containing probes can provide superior hybridization kinetics when the hybridization is conducted without carrier RNA (e.g., yeast RNA/yRNA) and in the presence of the nontoxic denaturant urea without compromising the specificity of the assays [13]. We demonstrated this by comparing the performance of DNA+LNA probes and 2OMe+LNA probes designed to target the same three microRNA targets (miR-9*, miR-130a, and miR-138), respectively. Using formalin-fixed paraffin-embedded (FFPE) sections of mouse brain tissue, we observed a tremendous effect of excluding the carrier RNA (yRNA) from the hybridization buffer when 2OMe+LNA was used compared to using DNA+LNA probes after a 1 h hybridization step (Fig. 1). In all the three cases we saw a significant increase in the hybridization signal obtained with the 2OMe+LNA probes in the absence of carrier RNA compared to using a hybridization buffer with carrier RNA (500 μg/mL), and very interestingly we did not see this effect when comparing DNA+LNA probes for the same targets. The images shown in Fig. 1 imply that we could not detect signal from any of the DNA+LNA probes; however, this is not true as prolonged exposure times indeed would reveal signal from these probes. But in order to clarify the difference in sensitivity of the different hybridization assays, the images in Fig. 1 were all acquired with identical exposure times for each target, respectively, determined by the condition of using 2OMe+LNA probes without yRNA.

We investigated the apparent inhibitory effect on the hybridization efficiency of 2OMe+LNA probes by carrier RNA. By quantitative comparison of miR-138 hybridizations performed for 1 or 3 h at different concentrations of yRNA we observed that increasing concentrations of yRNA gave a dose-dependent decrease in hybridization signal intensity both after 1 and 3 h of hybridization (Fig. 2), and we concluded that increasing hybridization times are likely to provide higher signal intensity, although in this case a much more pronounced effect on hybridization signal is obtained by excluding yRNA from the hybridization buffer [13].

Probe composition →	DNA + LNA		2OME-RNA + LNA	
Yeast RNA in hybr. buffer →	500 μg/mL	0 μg/mL	500 μg/mL	0 μg/mL
miR-9*				
miR-130a				
miR-138				

Fig. 1 In situ hybridizations of murine brain tissue sections to various miRNAs were performed for 1 h using DNA + LNA and 2OMe + LNA probes in hybridization buffers with or without yRNA. Detection was performed using a fluorescent TSA kit, with hybridization signal shown in red (Cy3) and cell nuclei are shown in blue through counterstaining with DAPI. Pictures for a specific miRNA (set of four images) were acquired at identical exposure settings to allow a visual comparison of hybridization signal intensities. The depicted regions of hybridization with miR-9* and miR-130a represent the cortex of the brain, and the region shown with miR-138 hybridization represents the cerebellum part of the mouse brain (Reprinted by Permission of SAGE Publication from Søe et al., *J Histochem Cytochem* **59(7)**, 661–672, © 2011 by The Authors

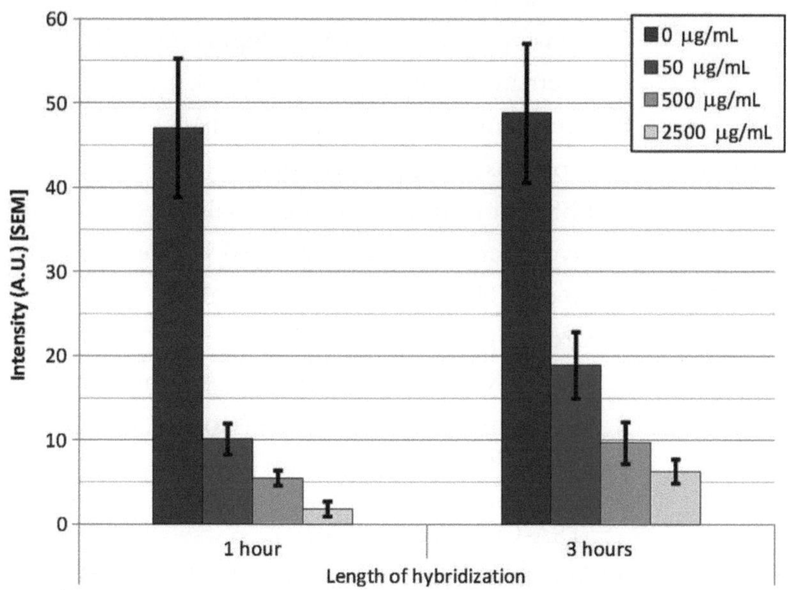

Fig. 2 Quantitative analysis of in situ hybridization signal intensity for detection of miR-138 using a 2OMe + LNA probe in presence of varying concentrations of yRNA. Hybridizations were performed for 1 or 3 h (Reprinted by Permission of SAGE Publication from Søe et al., *J Histochem Cytochem* **59(7)**, 661–72, © 2011 by The Authors)

It is important to note that the favorable effect of using 2OMe+LNA probes in hybridization buffers deprived of carrier RNA cannot be regarded as a general phenomenon of 2OMe-substituted LNA probes. We speculate that the relative abundance of a ncRNA target and potential differences in the affinity between probe and a given ncRNA target determines if a favorable effect of these probes compared to more commonly used DNA-substituted LNA-probes can be obtained [13]. We therefore suggest using this method especially for targets that prove to be difficult to detect or visualize using other techniques.

2 Materials

2.1 Special Reagents

1. Custom designed probes consisting of 2′-O-methyl RNA and LNA™ at every third position (RiboTask) specific for miR-138, miR-130a, and miR-9*. The probes were labeled with either digoxigenin (DIG) or fluorescein (FITC). Labeled control probes included a 2-base pair mismatch (2MM) probe for the miR-138 target and a scramble probe with no known targets. Commercially available miRCURY™ DNA+LNA™ probes (Exiqon) specific for miR-138, miR-130a, and miR-9*. The probes were 5′ labeled with either digoxigenin (DIG) or fluorescein (FITC).
2. RNase-depleted water, for example RNase-free Milli-Q water (Millipore).
3. Para-formaldehyde (PFA) (Sigma-Aldrich).
4. Triethanolamine (Sigma-Aldrich).
5. Acetic anhydride (Sigma-Aldrich).
6. Proteinase K (Sigma-Aldrich).
7. PBS 1× or 10×, RNase-free quality.
8. Hydrogen peroxide (H_2O_2) 30 % (Merck).
9. 20× SSC buffer, RNase-free quality.
10. Urea (Sigma-Aldrich).
11. Carrier RNA (yeast RNA/yRNA) (Ambion).
12. Sheep anti-fluorescein-POD (Roche). Sheep anti-digoxigenin-POD (Roche).
13. Tyramine signal amplification (TSA) Plus Cyanine 3 reagent (TSA-Cy3) (Perkin Elmer).
14. Anti-fade Prolong Gold with DAPI (Invitrogen).
15. For RNase-depleting working tools and surfaces: RNase ZAP, RNase Away or similar.

2.2 Buffers

1. *Post-fixation buffer*: to 45 mL Milli-Q water (preheated to 60 °C) add 20 μL 5 M NaOH and 2 g of paraformaldehyde and stir until dissolved. Add 5 mL 10× PBS. Cool solution on ice and adjust pH to 7.2. Filter the solution through a Whatman® qualitative filter paper, Grade 2.
2. *Acetylation buffer*: to 98 mL Milli-Q water, add 1.35 mL triethanolamine, 175 μL 12 M HCl, and 250 μL acetic anhydride.
3. *Proteinase K stock solution*: to 9.4 mL Milli-Q water, add 200 mg proteinase K, 200 μL of 1 M Tris–HCl (pH 7.4), 400 μL 1 M $CaCl_2$, and 10 mL glycerol. This gives a proteinase K concentration of 10 mg/mL. Make aliquots and store at −20 °C until use.
4. *Proteinase K reagent*: to 5 mL 1× PBS, add 12.5 μL proteinase-K stock of 10 mg/mL. This gives a final proteinase K concentration of 25 μg/mL (*see* **Note 1**).
5. *0.3% (v/v) H_2O_2 solution*: to 100 μL 30 % (v/v) H_2O_2 add 10 mL of methanol (*see* **Note 2**).
6. In situ *pre-hybridization buffer*: to 3.5 mL Milli-Q water, add 3.3 mL 12 M urea, 2.5 mL 20× SSC, 200 μL 50× Denhardt's solution, and 500 μL 10 mg/mL yRNA.
7. In situ *hybridization buffer*: to 4 mL Milli-Q water, add 3.3 mL 12 M urea, 2.5 mL 20× SSC, and 200 μL 50× Denhardt's solution.
8. *0.1× SSC buffer*: to 995 mL Milli-Q water, add 5 mL 20× SSC.
9. *PBS-T*: to 1 L of PBS (pH 7.4), add 1 mL Tween-20.
10. *Blocking solution*: to 38.5 mL of Milli-Q water, add 5 mL 1 M Tris–HCl (pH 7.5), 1.5 mL 5 M NaCl, 5 mL fetal bovine serum (FBS), and 40 μL 10 % Tween-20.

2.3 Glassware

In order to prevent RNase contamination, all glassware, including coplin jars, glass-staining racks, and stacks of coverslips and bottles for buffers, should be heat treated in an oven at 180 °C for 8 h. The items can be covered by aluminum foil before being placed in the oven in order to prevent contamination when removing the items afterwards.

3 Methods

3.1 Probe Design

MicroRNA target sequences were obtained from the miRBase homepage (www.mirbase.org). 2OMe + LNA probes were designed to be complementary to the full length of the following target miRNAs (miR-138, miR-130a, and miR-9*). LNA was used at the first position and then at every third position with 2OMe at the positions in between. Probes were 5′ labeled with either digoxigenin (DIG) or fluorescein (FITC). Labeled control probes included a 2-base pair mismatches (2MM) probe for the miR-138 target and a scramble probe with no known targets.

3.2 Preparation of Tissue Sections

Tissue sections should be prepared essentially as previously described [14]: Before cutting sections from the tissue blocks, the whole workstation (benchtop, microtome, blade holder, brushes, tweezers, cooling plate, water bath, etc.) should be cleaned with RNase ZAP. Set the cooling plate to −15 °C and place the FFPE block on the plate. Fill the water bath with RNase-free Milli-Q water and heat to 40–50 °C depending on the type of paraffin used for embedding. Another water bath is prepared as an RNase-free coplin jar containing room-temperature RNase-free Milli-Q water.

Insert a new disposable blade in the knife carrier, place the block in the cassette clamp. Trim the block, in order to avoid the first couple of sections, which might be contaminated during previous handling such as from tissue preparation or embedding. After trimming the block, cut 4 μm sections and place them into a dry sterilized jar with room-temperature RNase-free Milli-Q water, where any folding can be reversed. Transfer the slide to a heated (40–50 °C) water bath, where the tissue is stretched to avoid folds and overlaps in the structure and mount sections immediately on SuperFrost®Plus glass slides (Menzel-Gläser) (glass slides taken from a new and un-touched package are considered RNase-free). Mounting the section onto the slide should be done with care—allow water to slide away from in between the paraffin section and the glass slide—this process is important to avoid sections falling off during deparaffination. Let the slides dry for 2 h at room temperature and store at 4 °C in a dry box containing silica gel. Melt paraffin at 60 °C for 30 min on the day prior to the in situ hybridization experiment and store the slide at 4 °C.

3.3 Pretreatment of Tissue Slides

1. Deparaffinize slides in xylene and ethanol solutions in coplin jars. Place slides in xylene (two coplin jars) and then hydrate through ethanol solutions 99 % (two coplin jars), and 70 % (one coplin jar) to PBS (two coplin jars). Each solution should include 3-min incubation.
2. Post-fixation: treat the slides with 4 % PFA for 10 min.
3. Acetylate sections for 10 min in acetylation buffer.
4. De-masking of tissue sections is done by applying 300 μL/slide of the proteinase-K reagent at 25 μg/mL directly on the tissue and incubating for 40 min at 30 °C in a horizontal humidifying chamber (*see* **Note 1**).
5. Discard the proteinase-K reagent and wash three times with PBS.

6. Apply 300 μL freshly prepared 0.3 % H_2O_2 in methanol (*see* **Note 2**) and incubate at room temperature for 10 min.
7. Discard the H_2O_2 and wash three times with PBS.

3.4 In Situ Hybridization

1. Pre-hybridization: Tissue sections are incubated in pre-hybridization buffer (*see* **Note 3**) in a lab oven (*see* **Note 4**) for 30 min at the hybridization temperature (*see* **Note 5**) depending on the probe used in the hybridization step. In this case, for the miR-138 probes: 65 °C (2OMe+LNA) and 60 °C (DNA+LNA), for the miR-130a probes: 50 °C (both 2OMe+LNA and DNA+LNA), and for the miR-9* probes: 45 °C (both 2OMe+LNA and DNA+LNA).
2. Hybridization: Discard the pre-hybridization and add the hybridization buffer. The hybridization buffer is prepared by first, denaturing the probe and diluting it in the hybridization buffer. For example, for 2 mL hybridization mix containing 40 nM (*see* **Note 6**) DIG-labeled miR-138 2OMe+LNA probe (from a 25 μM probe stock), transfer 3.2 μL into the bottom of a 2-mL non-stick RNase-free tube and place the tube at 80 °C for 4 min. Spin down shortly using a table top centrifuge, and immediately add 2 mL hybridization buffer (preheated to the hybridization temperature) into the tube. Place the slides in a lab oven preset at the hybridization temperature (*see* **Note 4**). Incubate the slides for 1 h.
3. Place the slides into pre-warmed (determined by the hybridization temperature) 0.1×SSC in a coplin jar. The coverslips will easily detach. Transfer the slides to another casket with pre-warmed 0.1×SSC. Wash slides three times using pre-warmed 0.1×SSC.
4. Incubate with 200 μL blocking solution for 30 min at RT.
5. Wash the slides three times in PBS.
6. For detection of FITC-labeled probes, apply sheep anti-fluorescein-POD diluted 1:400 in blocking solution and incubate for 60 min at RT. To 1 mL blocking solution add 2.5 μL sheep anti-fluorescein-POD.
7. For detection of DIG-labeled probes, apply sheep anti-digoxigenin-POD diluted 1:400 in blocking solution and incubate for 60 min at RT. To 1 mL blocking solution add 2.5 μL sheep anti-digoxigenin-POD.
8. Wash each slide with 300 μL PBS-T three times for 3 min.
9. Incubate with 150 μL freshly prepared TSA-Cy3 reagent 7–15 min at RT. Protect from light during development. For the preparation of 1,500 μL TSA-Cy3 substrate, add 30 μL TSA-Cy3 reagent into 1,500 μL diluent.
10. Wash each slide with 300 μL PBS three times for 3 min.

11. Rinse the slides briefly with Milli-Q water and dry slides.
12. Mount the slides directly with Anti-fade Prolong Gold containing DAPI. Store the slides in dark at 4 °C.
13. Evaluate the slides using a fluorescence microscope with filters allowing detection of Cy3 and DAPI emission.

4 Notes

1. In this study we performed proteinase K treatment at 25 μg/mL at 30 °C for 40 min. The optimal proteinase-K treatment is dependent on the tissue to be analyzed and the fixation protocol used. Routinely processed tissue samples, which are usually fixed overnight at room temperature in neutral buffered formalin often show adequate performance. Denaturing fixatives should be avoided for use in ISH analyses. Proteinase K treatment efficiency is dependent on three conditional parameters: the proteinase K concentration, the incubation temperature, and the duration of incubation.
2. H_2O_2 is a potential harmful and corrosive oxidant reagent that should be handled with care. Methanol is an organic flammable and toxic solvent that should be handled under sufficient precautions.
3. The pre-hybridization buffer contains yRNA. The purpose of including yRNA in this step is to ensure blocking of nonspecific binding sites in the tissue sections for nucleic acid probes. Since the yRNA in this protocol is left out in the hybridization buffer it is important to note that the pre-hybridization step serves as a blocking step for nonspecific probe binding.
4. For the pre-hybridization and hybridization steps we used a conventional hybridization oven. We applied 50 μL of hybridization buffer and sealed with a coverslip glass.
5. The hybridization temperature should be chosen experimentally by evaluating the signal intensity obtained from an ISH using a specific probe compared to a scrambled or mismatch probe. Most often the hybridization temperature will be close to 60 °C, but can to the authors experience vary from 45 °C to 65 °C depending on the probe sequence. Hybridization temperatures below 45 °C will result in large nonspecific binding, while hybridization temperatures above 65 °C most often will result in very weak specific signals.
6. In this study we used the different probes at a similar concentration of 40 nM. The probe concentration range may vary from 10 to 100 nM. Attention should be paid to nonspecific binding and cross hybridization when incubating probes at high-range probe concentrations. In this study we used a negative

scrambled control probe which gives a good indication of non-specific probe binding, when incubated at the same concentration as the specific probe. Cross-hybridization, however, cannot be evaluated using the scrambled probe and requires other specificity controls such as mismatch controls. Here, we used in the case of miR-138 in situ hybridization a 2-base-pair mismatch (2MM) miR-138 probe.

Acknowledgments

The authors thank the Danish Ministry and Agency of Science, Technology, and Innovation for funding to MJS.

References

1. Kloosterman WP, Wienholds E, de Bruijn E (2006) In situ detection of miRNAs in animal embryos using LNA-modified oligonucleotide probes. Nat Methods 3:27–29
2. Silahtaroglu AN, Nolting D, Dyrskjøt L et al (2007) Detection of microRNAs in frozen tissue sections by fluorescence in situ hybridization using locked nucleic acid probes and tyramide signal amplification. Nat Protoc 2:2520–2528
3. Obernosterer G, Martinez J, Alenius M (2007) Locked nucleic acid-based in situ detection of microRNAs in mouse tissue sections. Nat Protoc 2:1508–1514
4. Nuovo GJ, Elton TS, Nana-Sinkam P et al (2009) A methodology for the combined in situ analyses of the precursor and mature forms of microRNAs and correlation with their putative targets. Nat Protoc 4:107–115
5. Pena JTG, Sohn-Lee C, Rouhanifard SH et al (2009) miRNA in situ hybridization in formaldehyde and EDC-fixed tissues. Nat Methods 6:139–141
6. Jørgensen S, Baker A, Møller S et al (2010) Robust one-day in situ hybridization protocol for detection of microRNAs in paraffin samples using LNA probes. Methods 52:375–381
7. Sempere LF, Preis M, Yezefski T et al (2010) Fluorescence-based codetection with protein markers reveals distinct cellular compartments for altered MicroRNA expression in solid tumors. Clin Cancer Res 16:4246–4255
8. Nielsen BS (2012) MicroRNA in situ hybridization. Methods Mol Biol 822:67–84
9. Renwick N, Cekan P, Masry PA et al (2013) Multicolor microRNA FISH effectively differentiates tumor types. J Clin Invest 123(6): 2694–2702. doi:pii: 68760. 10.1172/JCI68760
10. Koshkin AA, Singh SK, Nielsen P et al (1998) LNA (locked nucleic acids): synthesis of the adenine, cytosine, guanine, 5-methylcytosine, thymine and uracil bicyclonucleoside monomers, oligomerisation, and unprecedented nucleic acid recognition. Tetrahedron 54:3607–3630
11. Vester B, Wengel J (2004) LNA (locked nucleic acid): high-affinity targeting of complementary RNA and DNA. Biochemistry 43: 13233–13241
12. McTigue PM, Peterson RJ, Kahn JD (2004) Sequence-dependent thermodynamic parameters for locked nucleic acid (LNA)-DNA duplex formation. Biochemistry 43:5388–5405
13. Søe MJ, Møller T, Dufva M et al (2011) A sensitive alternative for microRNA in situ hybridizations using probes of 2′-O-methyl RNA+LNA. J Histochem Cytochem 59:661–672
14. Nielsen BS, Holmstrøm K (2013) Combined microRNA in situ hybridization and immunohistochemical detection of protein markers. Methods Mol Biol 986:353–365

Chapter 11

Enhanced Detection of Small RNAs Using a Nonradioactive Approach

Teresa T. Liu, Zhihua Li, and Bino John

Abstract

Recent advancements in high-throughput sequencing have led to the identification of many new classes of small noncoding RNAs such as endo-siRNAs. Unfortunately, reliable quantification of RNAs by sequencing is difficult due to artifacts arising from various factors involved in cDNA library preparation. Northern blot is one of the leading methods used to confirm the presence of a given RNA sequence because it can accurately quantify the cellular abundance, the size of the small RNA and reveal the presence of potential precursors and RNA isoforms. Here, we present a comprehensive description of LNA probe design along with a recently developed highly sensitive and cost-effective nonradioactive northern blot approach termed LED. LED combines a cross-linking method (EDC) and digoxigenin (DIG) labeling, and it can detect small RNAs with concentrations as low as 0.05 fmol and requires as little as a few seconds of membrane exposure for signal generation.

Key words Terminal modifications, 5′ phosphate, Membrane cross-linking, Locked nucleic acid probe, Nonradioactive detection

1 Introduction

Small noncoding RNAs are involved in gene regulation across a wide variety of organisms. These regulatory RNAs manifest a variety of complex regulatory modes that result in translational repression [1] and activation of both transcription [2] and translation [3]. Translational repression, the most common mode of small RNA-mediated regulation, generally occurs when the RNAs associate with the Argonaute (AGO) family of proteins that are contained within the RNA-induced silencing complex (RISC). Small RNAs such as microRNAs (miRNAs) and endo-siRNAs can guide RISC to destabilize mRNAs through endonuclease cleavage, deadenylation, and/or decapping, which can result in the blockage of translation initiation or protein elongation [4, 5]. Recent studies indicate that in at least some cases, the reduction in the rate of translation initiation precedes all other events such as deadenylation that ultimately affect

Andreas Werner (ed.), *Animal Endo-siRNAs: Methods and Protocols*, Methods in Molecular Biology, vol. 1173, DOI 10.1007/978-1-4939-0931-5_11, © Springer Science+Business Media New York 2014

mRNA stability [6, 7]. The diverse modes in which small RNAs function and the protein complexes shared by the various small RNA pathways suggest that these pathways represent complex regulatory cellular mechanisms that are highly intertwined.

Although many noncanonical small noncoding RNAs [8, 9] have been discovered in the recent past, the most widely studied small RNAs correspond to endo-siRNAs [10], miRNAs [1], and piRNAs [11–15]. Endo-siRNAs share disparate levels of similarities with both miRNAs and piRNAs in their biogenesis and function. Endo-siRNAs and miRNAs are ~20 nucleotides in length, processed by Dicer, and generally recognize their target transcripts by direct binding interactions with their targets. While targets of endo-siRNAs manifest near-perfect sequence complementarity to their target RNAs, miRNAs recognize their target mRNAs primarily via a seed sequence contained by the 5′ end of the miRNA [16]. Unlike miRNAs that originate from a single stranded primary RNA transcript (pri-miRNA) that forms a short hairpin structure and processed by Drosha, endo-siRNAs are processed independent of Drosha, from double-stranded RNAs (dsRNAs) or long hairpins. In general, endo-siRNAs are found across a wide variety of model organisms including plants, worms, flies, and mice. The endo-siRNAs function to posttranscriptionally silence transposons, viruses, and genes [10, 17–23]. In contrast, piRNAs are 24–31 nucleotides in length and are found mainly in germ cells. However, like endo-siRNAs, they function to silence the repetitive elements through transcriptional and posttranscriptional silencing mechanisms [12–15, 24, 25].

The detection of endo-siRNAs by sequencing and northern blots are helped by the fact they contain a 5′ monophosphate and a 2′,3′ hydroxyl at the 3′ end with a short overhang on both strands, a hallmark of Dicer-processing in animals [26, 27]. The unique presence of an uncapped 5′ end, along with the unmodified 3′ end allows adapter ligation and subsequent PCR amplification of these RNAs for high-throughput sequencing [28]. However, any modifications [29] present at either the 5′ or the 3′ end of the small RNAs could adversely affect small RNA detection methods that rely on 5′/3′ groups, particularly for sequencing-based methods that rely on adaptor ligation. In mammals, terminal modifications are known in RNAs such as piRNAs and secondary siRNAs that are processed independent of Dicer [5, 19, 30, 31]. For example, the RNA-dependent RNA polymerase (RdRP) generates secondary siRNAs in *C. elegans* containing 5′ polyphosphates [19]. The cloning and detection of such small RNAs requires the use of 5′ mono-phosphate independent protocols that relies on initial dephosphorylation followed by the phosphorylation of the 5′ end [19].

In this chapter we describe a highly sensitive, nonradioactive procedure to detect small RNAs [32]. The protocol (Fig. 1) takes advantage of the presence of a 5′ phosphate of the small RNA and

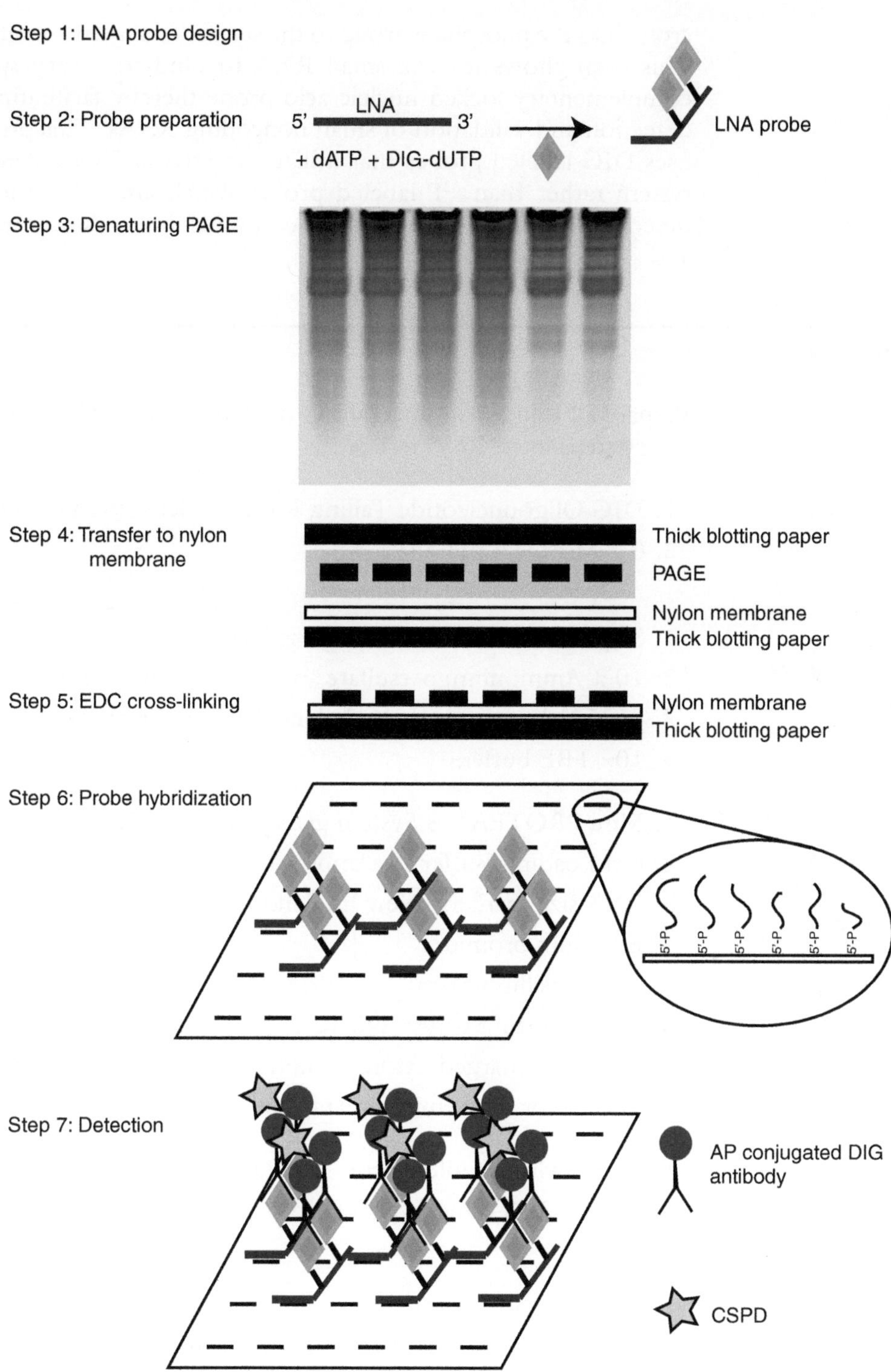

Fig. 1 Overview of the LED northern blot protocol. The LNA probe is designed (*Step 1*) and DIG labeled (*Step 2*). The RNA samples are prepared, electrophoresed (*Step 3*), transferred to a nylon membrane (*Step 4*) and EDC cross-linked (*Step 5*). DIG labeled probes are hybridized to the RNA (*Step 6*) and detected using an alkaline phosphatase (AP)-conjugated DIG antibody in conjunction with CSPD, a substrate of AP (*Step 7*)

cross-links the phosphate group to the surface of a nylon membrane. This then allows for the small RNA to bind to a very specific complementary locked nucleic acid probe thereby facilitating the detection and validation of small noncoding RNAs. This protocol uses DIG-labeled probes that utilize the DIG antibody detection system rather than ^{32}P-labeled probes which also allows for the detection of small noncoding RNAs at much lower concentrations and shorter exposure times.

2 Materials

Prepare all solutions using DEPC-treated water. Perform all laboratory steps in an RNA-free area.

2.1 LNA Probe Preparation Components

1. DIG Oligonucleotide Tailing Kit, 2nd Generation (Roche).
2. 0.2 M EDTA pH 8.0.

2.2 Urea polyacrylamide Gel Components

1. SequaGel Kit (National diagnostics).
2. 10 % Ammonium persulfate solution in water (*see* **Note 1**).
3. *N*,*N*,*N*,*N*′-tetramethyl-ethylenediamine (TEMED).
4. 10× TBE buffer.

2.3 Gel Electrophoresis and Electrotransfer Components

1. Mini PROTEAN 3 System glass plates (Bio-Rad; *see* **Note 2**).
2. Gel Loading Buffer II (Ambion).
3. RNA size markers (New England Biolabs).
4. Ethidium bromide.
5. UV transilluminator.
6. Thick blotting paper (Bio-Rad).
7. Positively charged nylon membrane (Roche; *see* **Note 3**).
8. Trans-Blot SD Semi-Dry Transfer Cell (Bio-Rad).

2.4 EDC Cross-linking Components

1. 1-Methylimidazole (Sigma, M50834).
2. 1-ethyl-3-(3-dimethylaminopropyl) carbodiimide (EDC; Sigma).

2.5 Hybridization and Detection Components

1. Hybridization bottles and oven.
2. ULTRAhyb Ultrasensitive Hybridization Buffer (Ambion; *see* **Note 4**).
3. Low Stringency Wash Buffer: 2× SSC with 0.1 % SDS.
4. High Stringency Wash Buffer: 0.1× SSC with 0.1 % SDS.
5. Wash Buffer: 1× SSC.
6. Anti-Digoxigenin-AP, Fab fragments (Roche).

7. CSPD (Roche).
8. DIG Wash and Block Buffer Set (Roche).
9. X-ray film or ChemiDoc-IT imaging system (UVP).

3 Methods

3.1 LNA Probe Design

1. Determine the reverse complement of the target small RNA. The candidate probe should be between 15 and 30 nucleotides (Fig. 2).
2. Add a stretch of five adenosine bases (polyA) to the end of the candidate probe sequence (*see* **Note 5**).
3. Enter candidate probe sequence into mFold to check for RNA secondary structure (*see* **Note 6**).
4. Manually shorten or lengthen probe so that mFold Tm ≤ 60 °C (*see* **Note 7**). The final sequence will be the probe backbone.
5. Transfer the backbone without the terminal As to the LNA Oligo Tm Prediction tool (http://www.exiqon.com/oligo-tools).
6. Manually change regular bases to LNA bases using plus signs in the LNA Oligo Tm Prediction tool until Tm ≥ 77 °C (*see* **Note 8**).
7. Transfer the resulting LNA probe sequence to the LNA Oligo Optimizer tool and calculate both secondary structure and

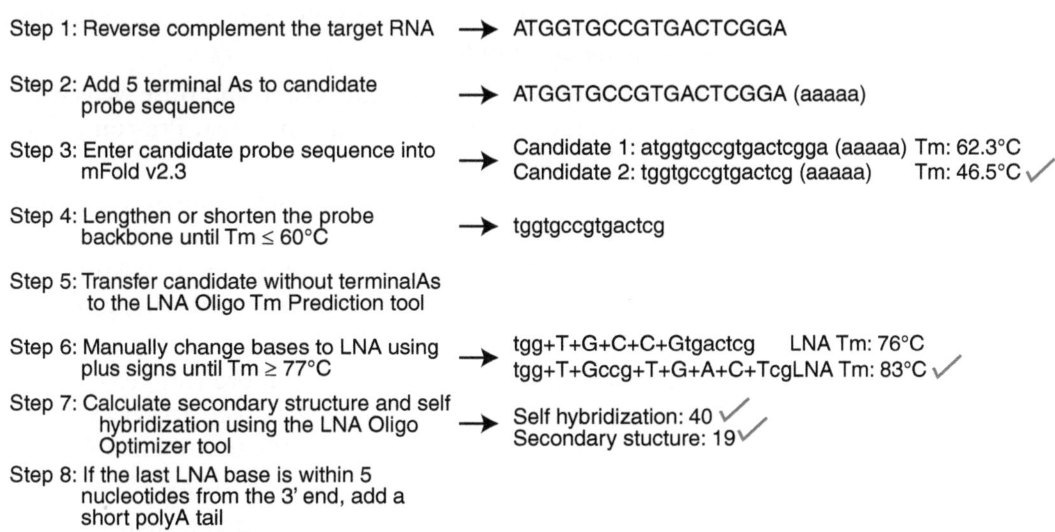

Fig. 2 Detailed workflow for designing effective LNA probes

self-hybridization. The secondary structure score should be ≤25 and the self-hybridization score should be ≤45. Both scores should be as low as possible to minimize self-hybridization of the probe.

8. If the last LNA base is less than 5 nucleotides away from the 3′ end, add a short polyA tail to LNA probe so that the last LNA base is at least 5 nucleotides from the 3′ end (*see* **Note 9**).
9. If compromises must be made in **steps 6** and 7, the priority is as follows: self-hybridization score can be increased up to 50, secondary structure score can be increased up to 30, and the LNA Tm can be lowered to 75 °C. The mFold Tm must strictly be ≤60 °C (*see* **Note 10**).

3.2 Probe Preparation with dATP/DIG-dUTP Tailing

1. In a 1.5 mL tube kept on ice, add 2 μL 50 pmol LNA probe to 7 μL DEPC-treated water (*see* **Note 11**).
2. Add, in order, 4 μL reaction buffer (vial 1), 4 μL $CoCl_2$-solution (vial 2), 1 μL DIG-dUTP solution (vial 3), 1 μL dATP solution (vial 4), and 1 μL 400 U Terminal transferase (vial 5) to a tube containing the LNA probe solution. Mix and centrifuge briefly.
3. Incubate at 37 °C for 15 min and then place on ice.
4. Stop the reaction by adding 2 μL 0.2 M EDTA pH 8.0.
5. Store probe at −20 °C (*see* **Note 12**).

3.3 15 % Denaturing PAGE

1. Mix 6 mL SequaGel Concentrate, 3 mL SequaGel Diluent, 1 mL SequaGel Buffer, 4 μL TEMED, and 40 μL 10 % ammonium persulfate (*see* **Note 13**). Cast gel all the way to the top of 10.1 × 7.3 cm^2 plates with 1.5 mm spacers cleaned carefully with DEPC-treated water. Carefully insert a gel comb without introducing air bubbles.
2. After gel polymerization, remove comb and rinse wells with 1× TBE to remove excess acrylamide and urea. Pre-run gel in 1× TBE at ~200–400 V for 1 h.
3. Add Gel Loading Buffer II to RNA samples (~5–20 μg). Denature the RNA samples at 95 °C for 5 min and then chill on ice. Prepare RNA markers according to manufacturer specifications and add ethidium bromide at 10–50 μg/mL final concentration.
4. Prior to loading samples, rinse wells of pre-run gels to remove acrylamide debris. Load samples and markers to gel and electrophorese at 200 V for 50 min or until the bromophenol blue dye reaches about 1 cm above the bottom of the gel (*see* **Note 14**).
5. Following electrophoresis, disassemble the gel plates by gently prying the plates apart. Image the gel with a transilluminator to visualize the RNA size markers. Soak the gel in 1× TBE for 10 min with gentle rocking.

3.4 Semi-Dry Electrophoretic Transfer

1. Cut nylon membrane and two sheets of thick blot paper to the size of the gel. Wet membrane briefly with DEPC-treated water and incubate with 1× TBE for 5 min with gentle rocking. Briefly wet the thick blot paper in 1× TBE.
2. Assemble the gel sandwich on the positive electrode of the semi-dry blotter as follows: blot paper, nylon membrane, gel, blot paper (*see* **Note 15**).
3. Transfer RNA to membrane at 380 mA (~10–15 V) for 90 min.

3.5 EDC Cross-linking

1. Add 245 μL 12.5 M 1-methylimidazole to 9 mL DEPC-treated water. Add 300 μL 1 M HCl to maintain a pH of 8.0 (*see* **Note 16**).
2. Immediately before use, add 0.753 g EDC to the 1-methylimidazole solution from **step 1** and bring the volume up to 24 mL with DEPC-treated water. This is the EDC cross-linking solution.
3. Add cross-linking solution to one sheet of thick blot paper. Place nylon membrane RNA side up on blot paper soaked with EDC cross-linking solution. Wrap the blot paper and membrane in plastic wrap and incubate with the blot paper side down on a heat block at 60 °C for 1–2 h.
4. Remove excess cross-linking solution with DEPC-treated water (*see* **Note 17**).

3.6 LNA Probe Hybridization

1. Preheat hybridization buffer to 37 °C until all precipitated materials have been resuspended.
2. After cross-linking, place membrane and 15 mL ULTRAhyb hybridization buffer into hybridization bottle (*see* **Note 18**).
3. Pre-hybridize the membrane at 37 °C for at least 30 min in a hybridization oven fitted with a rotisserie.
4. Denature the probe at 95 °C for 1 min and add probe to 15 mL fresh ULTRAhyb buffer to a final concentration of 0.5 nM.
5. Discard pre-hybridization buffer and replace with hybridization buffer plus probe solution. Incubate with slow rotation at 37 °C overnight in hybridization oven.

3.7 Blocking and Detection

1. Preheat washing buffers at 37 °C. Discard hybridization buffer plus probe solution from bottle.
2. Wash the membrane twice with Low Stringency Buffer at 37 °C for 15 min.
3. Wash the membrane twice with High Stringency Buffer at 37 °C for 15 min.
4. Incubate the membrane with Wash Buffer at 37 °C for 15 min.

5. Dilute 10× Blocking Solution with 1× Maleic acid buffer to a 1× Blocking Solution. Remove the membrane from hybridization bottle and incubate the membrane in Blocking Solution for 3 h at room temperature.
6. Dilute anti-DIG antibody 1:15,000 in 1× Blocking Solution. Discard Blocking Solution and incubate the membrane with DIG antibody Blocking Solution for 30 min at room temperature.
7. Dilute 10x DIG Washing Buffer to 1× solution with DEPC-treated water. Wash the membrane four times for 15 min each at room temperature.
8. Dilute 10× DIG Development Buffer to 1× solution with DEPC-treated water. Incubate the membrane in 1× DIG Development Buffer for 5 min at room temperature.
9. Dilute CSPD 1:100 in 1× DIG Development Buffer to make the Substrate Solution. Discard 1× DIG Development Buffer and replace with the Substrate Solution. Incubate for 5 min at room temperature (*see* **Note 19**).
10. Carefully remove Substrate Solution by touching the corner of the membrane to a clean paper towel to wick away excess buffer. Wrap in clean plastic wrap and expose blot to X-ray film or detect using a ChemiDoc imager.

4 Notes

1. Ammonium persulfate (10 %) solution can be stored at −20 °C in aliquots, for future use.
2. Any vertical protein gel electrophoresis system would be sufficient.
3. Four different nylon membranes were tested [32] and the positively charged nylon membrane from Roche or the neutral nylon membrane from GE Healthcare yielded the strongest signal (Fig. 3a).
4. Various hybridization buffers were tested [32]; the Ambion ULTRAHyb was able to detect both highly abundant and less abundant miRNAs at the lowest hybridization temperature of 37 °C, thereby reducing required melting temperature of the probe–target hybridization. However, the DIG Easy Hyb was much more sensitive in the detection of miR-16 but requiring a higher hybridization temperature of 50 °C (Fig. 3b).
5. The polyadenylated form of the candidate probe sequence is used to calculate the Tm because probes are subjected to dATP/DIG-dUTP tailing during probe preparation.
6. mFold can be accessed at http://mfold.rna.albany.edu/?q=mfold/RNA-Folding-Form. Version 2.3 of mFold can calculate the Tm of the secondary structure of the RNA probe.

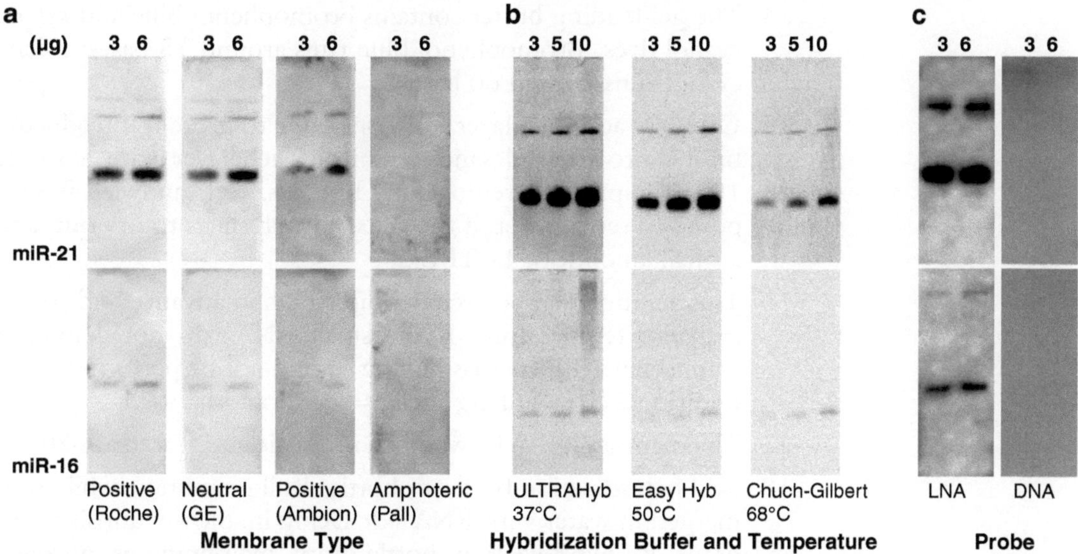

Fig. 3 Effects of different northern blot reagents on LED sensitivity in detecting small RNAs. Total RNA from MCF-7 cells was used to examine the presence of miR-21 and miR-16 using LNA probes unless otherwise indicated. (**a**) Performance comparisons of four different nylon membranes. (**b**) Comparison of four different hybridization buffers at their optimum hybridization temperatures. (**c**) Effect of LNA and DNA oligo probes on detection sensitivities for the tested small RNAs

7. The lower the Tm of the probe backbone, the less likely the probe backbone will form secondary structures leading to better hybridization results.
8. Start with 8 LNA bases in the middle of the backbone. The higher the LNA Oligo Tm, the less the nonspecific hybridizations will remain after the stringency washes.
9. The addition of dummy nucleotides at the 3′ end of the probe ensures that the terminal transferase recognizes the 3′ end during dATP/DIG-dUTP tailing. However, this final LNA probe sequence with the dummy nucleotides does not need to be examined with the LNA Oligo Tm predictor and Optimizer tools.
10. The lower self-hybridization and secondary structure scores result in better probe–target hybridization. If the LNA Oligo Tm is lowered to 75 °C, the probe may or may not detect the target sequence.
11. DNA probes may also be used instead of LNA probes if the small RNA is abundant, but no bands were detected using the DNA probe (Fig. 3c).
12. The DIG labeled LNA probe can be stored at −20 °C for up to 3 months.
13. Decreasing the acrylamide percentage gel from 15 % could result in poor resolution of small RNA bands.

14. The gel loading buffer contains bromophenol blue and xylene cyanol dyes. Bromophenol blue runs around 15 bases; xylene cyanol runs around 60 bases.
15. Carefully add each layer from one edge to prevent introducing bubbles to the gel sandwich. Roll each layer with a sterile 10 mL pipette to remove any bubbles. Gel sandwich is supposed to remain wet; if the gel sandwich starts to dry out, add a small amount of 1× TBE.
16. This solution can be prepared up to 2 h in advance and stored at room temperature. If the small RNAs do not contain a 5′-phoshate, EDC cross-linking should not be performed, and traditional cross-linking such as UV should be used.
17. The membrane can be wrapped in plastic and stored at −20 °C.
18. Add hybridization buffer to bottle. Roll membrane while submerged in water with RNA side facing in. Slowly unroll membrane in hybridization bottle until membrane is unrolled against the wall of the bottle. Avoid introducing bubble between the bottle and the membrane to prevent the membrane from drying out.
19. The membrane can be placed on a clean plastic wrap surface and the Substrate Solution applied to the surface without agitation.

References

1. Lee RC, Feinbaum RL, Ambros V (1993) The C. elegans heterochronic gene lin-4 encodes small RNAs with antisense complementarity to lin-14. Cell 75:843–854
2. Li LC, Okino ST, Zhao H, Pookot D, Place RF, Urakami S, Enokida H, Dahiya R (2006) Small dsRNAs induce transcriptional activation in human cells. Proc Natl Acad Sci U S A 103: 17337–17342
3. Vasudevan S, Tong Y, Steitz JA (2007) Switching from repression to activation: microRNAs can up-regulate translation. Science 318:1931–1934
4. Farazi TA, Juranek SA, Tuschl T (2008) The growing catalog of small RNAs and their association with distinct Argonaute/Piwi family members. Development 135:1201–1214
5. Kim VN, Han J, Siomi MC (2009) Biogenesis of small RNAs in animals. Nat Rev Mol Cell Biol 10:126–139
6. Bazzini AA, Lee MT, Giraldez AJ (2012) Ribosome profiling shows that miR-430 reduces translation before causing mRNA decay in zebrafish. Science 336:233–237
7. Djuranovic S, Nahvi A, Green R (2012) miRNA-mediated gene silencing by translational repression followed by mRNA deadenylation and decay. Science 336:237–240
8. Tuck AC, Tollervey D (2011) RNA in pieces. Trends Genet 27:422–432
9. Jacquier A (2009) The complex eukaryotic transcriptome: unexpected pervasive transcription and novel small RNAs. Nat Rev Genet 10:833–844
10. Hamilton A, Voinnet O, Chappell L, Baulcombe D (2002) Two classes of short interfering RNA in RNA silencing. EMBO J 21:4671–4679
11. Ruby JG, Jan C, Player C, Axtell MJ, Lee W, Nusbaum C, Ge H, Bartel DP (2006) Large-scale sequencing reveals 21U-RNAs and additional MicroRNAs and endogenousk siRNAs in C. elegans. Cell 127:1193–1207
12. Girard A, Sachidanandam R, Hannon GJ, Carmell MA (2006) A germline-specific class of small RNAs binds mammalian Piwi proteins. Nature 442:199–202
13. Aravin A, Gaidatzis D, Pfeffer S, Lagos-Quintana M, Landgraf P, Iovino N, Morris P, Brownstein MJ, Kuramochi-Miyagawa S, Nakano T, Chien M, Russo JJ, Ju J, Sheridan R, Sander C, Zavolan M, Tuschl T (2006) A

novel class of small RNAs bind to MILI protein in mouse testes. Nature 442:203–207

14. Lau NC, Seto AG, Kim J, Kuramochi-Miyagawa S, Nakano T, Bartel DP, Kingston RE (2006) Characterization of the piRNA complex from rat testes. Science 313:363–367
15. Vagin VV, Sigova A, Li C, Seitz H, Gvozdev V, Zamore PD (2006) A distinct small RNA pathway silences selfish genetic elements in the germline. Science 313:320–324
16. Bartel DP (2009) MicroRNAs: target recognition and regulatory functions. Cell 136: 215–233
17. Ghildiyal M, Seitz H, Horwich MD, Li C, Du T, Lee S, Xu J, Kittler EL, Zapp ML, Weng Z, Zamore PD (2008) Endogenous siRNAs derived from transposons and mRNAs in Drosophila somatic cells. Science 320:1077–1081
18. Vazquez F, Vaucheret H, Rajagopalan R, Lepers C, Gasciolli V, Mallory AC, Hilbert JL, Bartel DP, Crete P (2004) Endogenous trans-acting siRNAs regulate the accumulation of Arabidopsis mRNAs. Mol Cell 16:69–79
19. Pak J, Fire A (2007) Distinct populations of primary and secondary effectors during RNAi in C. elegans. Science 315:241–244
20. Chung WJ, Okamura K, Martin R, Lai EC (2008) Endogenous RNA interference provides a somatic defense against Drosophila transposons. Curr Biol 18:795–802
21. Czech B, Malone CD, Zhou R, Stark A, Schlingeheyde C, Dus M, Perrimon N, Kellis M, Wohlschlegel JA, Sachidanandam R, Hannon GJ, Brennecke J (2008) An endogenous small interfering RNA pathway in Drosophila. Nature 453:798–802
22. Tam OH, Aravin AA, Stein P, Girard A, Murchison EP, Cheloufi S, Hodges E, Anger M, Sachidanandam R, Schultz RM, Hannon GJ (2008) Pseudogene-derived small interfering RNAs regulate gene expression in mouse oocytes. Nature 453:534–538
23. Katiyar-Agarwal S, Morgan R, Dahlbeck D, Borsani O, Villegas A Jr, Zhu JK, Staskawicz BJ, Jin H (2006) A pathogen-inducible endogenous siRNA in plant immunity. Proc Natl Acad Sci U S A 103:18002–18007
24. Houwing S, Kamminga LM, Berezikov E, Cronembold D, Girard A, van den Elst H, Filippov DV, Blaser H, Raz E, Moens CB, Plasterk RH, Hannon GJ, Draper BW, Ketting RF (2007) A role for Piwi and piRNAs in germ cell maintenance and transposon silencing in Zebrafish. Cell 129:69–82
25. Shpiz S, Olovnikov I, Sergeeva A, Lavrov S, Abramov Y, Savitsky M, Kalmykova A (2011) Mechanism of the piRNA-mediated silencing of Drosophila telomeric retrotransposons. Nucleic Acids Res 39:8703–8711
26. Lee Y, Ahn C, Han J, Choi H, Kim J, Yim J, Lee J, Provost P, Radmark O, Kim S, Kim VN (2003) The nuclear RNase III Drosha initiates microRNA processing. Nature 425:415–419
27. Hammond SM, Bernstein E, Beach D, Hannon GJ (2000) An RNA-directed nuclease mediates post-transcriptional gene silencing in Drosophila cells. Nature 404:293–296
28. Hafner M, Renwick N, Farazi TA, Mihailovic A, Pena JT, Tuschl T (2012) Barcoded cDNA library preparation for small RNA profiling by next-generation sequencing. Methods 58: 164–170
29. Kim YK, Heo I, Kim VN (2010) Modifications of small RNAs and their associated proteins. Cell 143:703–709
30. Cifuentes D, Xue H, Taylor DW, Patnode H, Mishima Y, Cheloufi S, Ma E, Mane S, Hannon GJ, Lawson ND, Wolfe SA, Giraldez AJ (2010) A novel miRNA processing pathway independent of Dicer requires Argonaute2 catalytic activity. Science 328:1694–1698
31. Li Z, Ender C, Meister G, Moore PS, Chang Y, John B (2012) Extensive terminal and asymmetric processing of small RNAs from rRNAs, snoRNAs, snRNAs, and tRNAs. Nucleic Acids Res 40:6787–6799
32. Kim SW, Li Z, Moore PS, Monaghan AP, Chang Y, Nichols M, John B (2010) A sensitive non-radioactive northern blot method to detect small RNAs. Nucleic Acids Res 38:e98

Chapter 12

Computing siRNA and piRNA Overlap Signatures

Christophe Antoniewski

Abstract

High-throughput sequencing approaches opened the possibility to precisely map full populations of small RNAs to the genomic loci from which they originate. A bioinformatic approach revealed a strong tendency of sense and antisense piRNAs to overlap with each other over ten nucleotides and had a major role in understanding the mechanisms of piRNA biogenesis. Using similar approaches, it is possible to detect a tendency of sense and antisense siRNAs to overlap over 19 nucleotides. Thus, the so-called overlap signature which describes the tendency of small RNA to map in a specific way relative to each other has become the approach of choice to identify and characterize specific classes of small RNAs. Although simple in essence, the bioinformatic methods used for this approach are not easily accessible to biologists. Here we provide a python software that can be run on most of desktop or laptop computers to compute small RNA signatures from files of sequencing read alignments. Moreover, we describe and illustrate step by step two different algorithms at the core of the software and which were previously used in a number of works.

Key words High-throughput sequencing, Small RNA signature, siRNA, piRNA

1 Introduction

PIWI-interacting RNAs (piRNAs) are a major class of gonad-specific, ~24–29 nt long small RNAs involved in transposons and retrotransposons silencing, both in vertebrates and invertebrates. Their biogenesis has long remained enigmatic, as, in contrast to siRNA biogenesis, it does not involve the activity of Dicer enzymes. In 2007, two seminal works revealed that the production of piRNAs in *Drosophila* germ cells requires an amplification mechanism called the "Ping-Pong cycle" [1, 2]. In the Ping-Pong cycle, an antisense piRNA guides an Argonaute protein of the PIWI subfamily (Aubergine or Piwi) to the cleavage of a complementary transposon sense transcript. One of the two cleavage products gives rise to a sense piRNA–Agonaute-3 complex which can in turn guide the cleavage of a complementary transposon antisense transcript, generating a new antisense piRNA. Because all three

Andreas Werner (ed.), *Animal Endo-siRNAs: Methods and Protocols*, Methods in Molecular Biology, vol. 1173, DOI 10.1007/978-1-4939-0931-5_12,

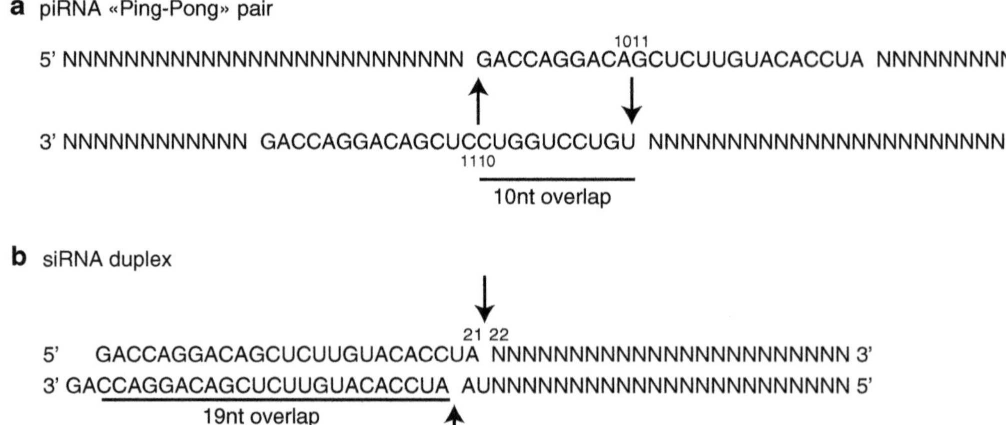

Fig. 1 piRNA "Ping-Pong" pairs and siRNA duplexes. (**a**) A typical pair of sense and antisense piRNAs overlapping by 10 nt. The 5′ end of piRNAs is determined by Argonaute-mediated cleavage (*arrows*) of a long piRNA precursor transcript (Ns) that occurs between positions 10 and 11 relative the 5′ end of the complementary guide piRNA. (**b**) An siRNA duplex with typical 3′ 2 nt overhang and 19 nt overlap between sense and antisense siRNAs is produced by cleavage of long double-stranded RNAs by the type III RNAse Dicer (*arrows*)

Drosophila PIWI Argonaute proteins cleave their target RNAs between nucleotides 10 and 11 relative to the 5′ end of their piRNA guides, iterations of the "Ping-Pong" mechanism lead to the accumulation of pairs of sense and antisense piRNAs which overlap exactly by ten nucleotides (Fig. 1a).

Figuratively speaking, high-throughput sequencing (HTS) technologies had an explosive impact on the field of piRNA biology. HTS of small RNAs allowed uncovering the specific features (sizes and nucleotide compositions) of piRNAs bound to Piwi, Aubergine, or Agonaute-3. Moreover, alignment software adapted to HTS datasets gave access to piRNA genome mapping at nucleotide level and allowed measuring, through computing approaches, their striking tendency to form 10 nt-overlapping pairs [2–8]. Interestingly, these simple but elegant statistical approaches to search for the so-called piRNA signature can also apply to siRNAs. Accordingly, if the Ping-Pong mechanism for piRNA production leaves a 10 nt-overlap signature in sequencing datasets, generation of siRNA duplexes by Dicer enzymes is expected to leave a 19 nt-overlap signature (Fig. 1b).

In this chapter, we describe two similar but distinct algorithms first used for the detection of the piRNA Ping-Pong signature [2, 4]. The algorithms were also able to detect 19 nt-overlap "Dicer" signatures in HTS data from siRNAs co-immunoprecipitated with the p19 viral suppressor of RNAi (*see* ref. [9], and Chapter 13 in this book) and, more recently, in sequencing data of viral siRNAs generated during *Drosophila* Nora virus infections [10]. We provide a python language script that implements these algorithms and can easily be used by biologists without advanced skills in bioinformatics to search for small RNA signatures with the help of any desktop computer equipped with a python interpreter.

2 Materials

A computer with at least 4 Gb of RAM.

Basic knowledge in UNIX management. Being able to open a UNIX terminal, to navigate in the file system and to launch simple command lines is sufficient.

Python interpreter installed. Python is pre-installed in Linux and Mac OS X standard configurations and can be easily installed on Window computers (http://www.python.org/getit/windows/).

Collections and *numpy* python modules installed. If this is not the case (unlikely), see how to install the modules at http://docs.python.org/.

An alignment file reporting small RNA sequence alignment to a reference genome. The script takes as input standard SAM files (http://samtools.sourceforge.net/SAM1.pdf). It can also take a more compact tabular format generated by bowtie (*see* ref. 11 and **Note 1**) with one reported match by line and (1) the read identifier in the first column, (2) the matched reference strand (+ or –) in the second column, (3) the name of the matched region in the reference genome (which corresponds to the sequence header when the reference genome is in a fasta format), (4) the 5′ coordinate of the match (0-based leftmost mapping position) always given relative to the sense strand (+) of the genome reference, and (5) the sequence of the matched read. It is important that the alignment file only reports matches to genomic targets relevant to the investigation. For example, searching for signatures of Nora virus derived siRNAs should not be performed using an input file that reports alignments to both *Drosophila* and Nora virus genomes (*see* **Note 2**).

2.1 Installation

1. Create a “test” directory.
2. Download the *signature.py* script file at http://drosophile.org/GEDlab/?page_id=730, or copy/paste the script in a text file named *signature.py* (*see* Appendix 1). In python language, code indentation is critical. If you copy/paste the signature.py code in a new file, make sure that the numbers of space characters at the beginning of lines have not been modified. Put a copy of *signature.py* in your test directory.

3. Make the *signature.py* script executable. This step is optional (*see* **Note 3**). Under UNIX compliant OS, it can be performed by navigating to your test directory (*cd ~/test*), typing *chmod 755 signature.py*, and pressing the enter key.

2.2 Running Signature.py

The script is run using the command line:

signature.py <input_file> <min_size> <max_size> <min_overlap> <max_overlap> <output_file>

where *<input_file>* is the path to the sequence alignment file (in a SAM or a bowtie format), *<min_size>* is the minimal size (integer) of small RNA reads to be analyzed, *<max_size>* is the maximal size (integer) of small RNA reads to be analyzed, *<min_overlap>* is the minimum overlap searched for small RNA pairs (an integer that expresses a 1-based overlap; a value of 1 means that the 5′ ends of small RNA pairs overlap with one nucleotide), *<max_overlap>* is the maximum overlap searched for small RNA pairs (an integer that expresses a 1-base overlap; a value of 21 will search for small RNA overlapping up to 21 nt, *see* **Note 3**), and <output_file> is the path to the output text file generated by the script.

Example: *signature.py bowtie_viral_matches.sam 21 21 1 21 myresults.txt* analyzes the overlapping tendencies of 21 nt reads matched in the *bowtie_viral_matches.sam* and outputs a result data frame in the *myresults.txt* file.

3 Methods

Here we describe in greater details the operations performed by the signature.py script. This section will help researchers with python expertise to adapt or extend the script to accommodate specific needs. However, we tried to simplify it as far as possible to help biologists to understand the two algorithms used for finding small RNA overlap signatures so that they are able to evaluate the biological significance of their analyses. Note that the terms "instance," "class," and "method" used below are typical python terms related to Object Oriented Programming.

3.1 Data Treatment by Signature.py Python Script

The signature.py script reads tabular bowtie or SAM alignment reports and implements as many instances of the *smRNAwindow* python class (*see* **Note 4**) as necessary. It is noteworthy that an *smRNAwindow* instance is generated for each distinct item found in the alignment file (by item, we mean genes, chromosomes, viral genomes or subgenomes, etc.). For example, if the alignment file contains only matches to a single molecule RNA virus genome, python will generate a single *smRNAwindow* instance. In contrast, if the alignment file contains matches to an RNA virus genome with two subgenomic A and B fragments, python will generate two *smRNAwindow* instances. Along the same line, if the alignment file reports matches to *Drosophila* genes, python will generate ~18,000 *smRNAwindow* instances.

3.2 The smRNAwindow Class

In addition to the standard *def __init__* python method which instantiates a python dictionary, the *smRNAwindow* class has four methods (*see* **Note 5**).

The *addread* method adds reads matched to an item to the corresponding *smRNAwindow* instance, which is indexed with this item in the *objDic* dictionary of *smRNAwindow* instances. Note that *smRNAwindow* instances are created when necessary on the fly during the initial data acquisition process.

The *readcount* method computes the number of reads of minimum and maximum sizes (given as method parameters) attributed to an *smRNAwindow* instance.

The *count_pairs* method computes the number of small RNA pairs found in an *smRNAwindow* instance, with an overlap of *minscope* to *maxscope* nucleotides (the "scope" of the signature). When *count_pairs* is called, each position of small RNAs with the appropriate size (defined by the minsize and maxsize parameters) is iteratively tested to find whether reads are referenced to an overlapping position of the opposite strand, within the *scope* range. When this happens, the lower number of reads between the query position and the "paired" position is added to a local histogram of number of pairs by class of overlap values. For instance (Fig 2a), if 9 reads at the query position *p* on the forward strand are found to overlap 2 reads at the position *p+19* (19 nt overlap) on the reverse strand, the class 10 of the local histogram is incremented by 2 (only two pairs can be formed in this example). Along the iteration process, local histograms for each query position are summed. As all positions of both forward and reverse strands are examined, pairs of small RNA reads are counted twice. Therefore, the values in the summed histogram are divided by 2, before being returned by the *count_pairs* method, as a histogram of number of pairs by class of overlap values (Fig. 2b).

The *overlap_probability* method computes the probability, within a *smRNAwindow* instance, to find an antisense read overlapping with a sense read over *minscope* to *maxscope* nucleotides (the "scope" of the signature). The *overlap_probability* method of the *smRNAwindow* python class is inspired from the procedure first described by Brennecke et al. [2] and further formalized in ([3], *see* supporting online material of this reference).

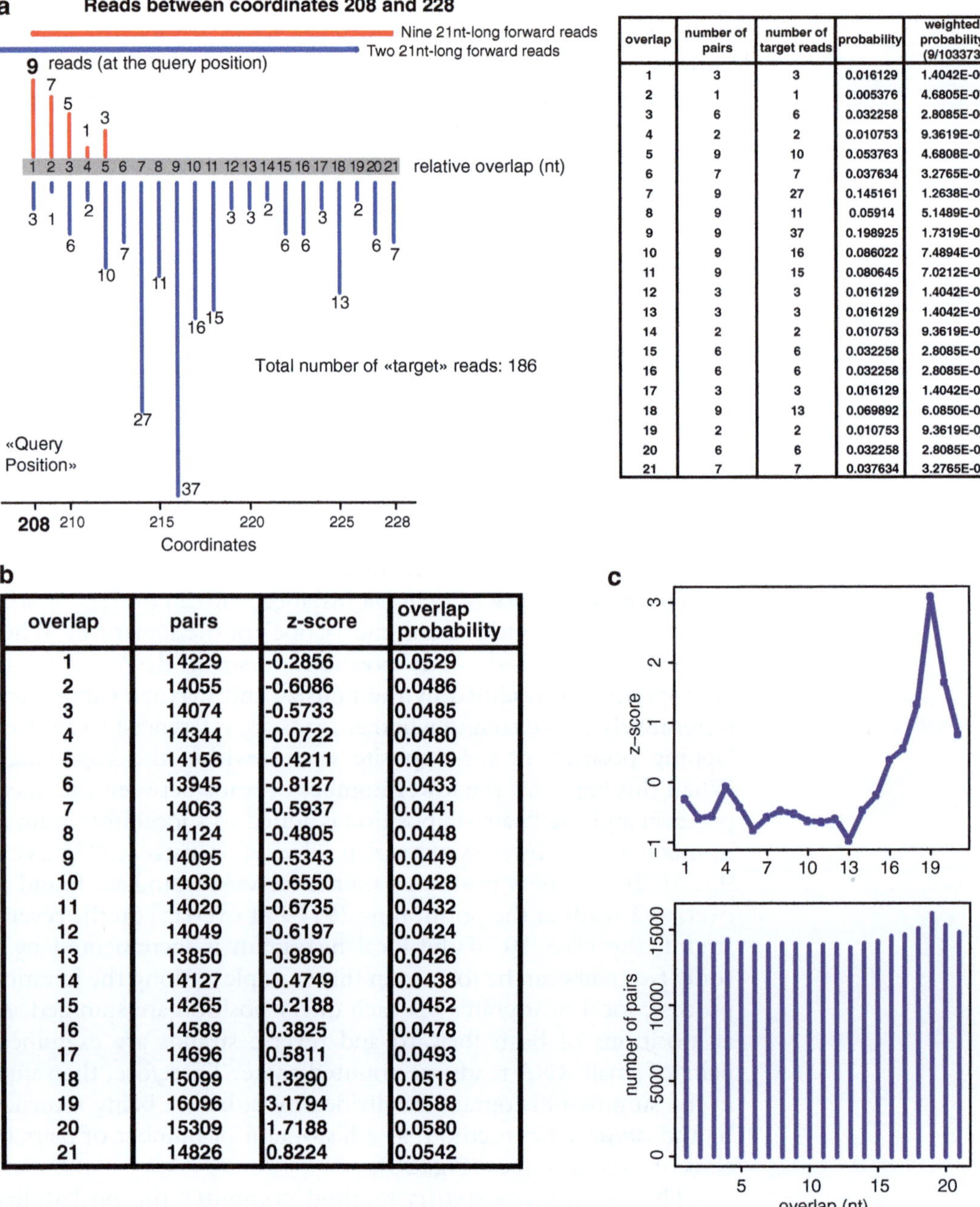

overlap	number of pairs	number of target reads	probability	weighted probability (9/103373)
1	3	3	0.016129	1.4042E-06
2	1	1	0.005376	4.6805E-07
3	6	6	0.032258	2.8085E-06
4	2	2	0.010753	9.3619E-07
5	9	10	0.053763	4.6808E-06
6	7	7	0.037634	3.2765E-06
7	9	27	0.145161	1.2638E-05
8	9	11	0.05914	5.1489E-06
9	9	37	0.198925	1.7319E-05
10	9	16	0.086022	7.4894E-06
11	9	15	0.080645	7.0212E-06
12	3	3	0.016129	1.4042E-06
13	3	3	0.016129	1.4042E-06
14	2	2	0.010753	9.3619E-07
15	6	6	0.032258	2.8085E-06
16	6	6	0.032258	2.8085E-06
17	3	3	0.016129	1.4042E-06
18	9	13	0.069892	6.0850E-06
19	2	2	0.010753	9.3619E-07
20	6	6	0.032258	2.8085E-06
21	7	7	0.037634	3.2765E-06

overlap	pairs	z-score	overlap probability
1	14229	-0.2856	0.0529
2	14055	-0.6086	0.0486
3	14074	-0.5733	0.0485
4	14344	-0.0722	0.0480
5	14156	-0.4211	0.0449
6	13945	-0.8127	0.0433
7	14063	-0.5937	0.0441
8	14124	-0.4805	0.0448
9	14095	-0.5343	0.0449
10	14030	-0.6550	0.0428
11	14020	-0.6735	0.0432
12	14049	-0.6197	0.0424
13	13850	-0.9890	0.0426
14	14127	-0.4749	0.0438
15	14265	-0.2188	0.0452
16	14589	0.3825	0.0478
17	14696	0.5811	0.0493
18	15099	1.3290	0.0518
19	16096	3.1794	0.0588
20	15309	1.7188	0.0580
21	14826	0.8224	0.0542

Fig. 2 Local and global histograms of number of pairs and of overlap probabilities. (**a**) An example of local histograms computed by signature.py at the query position 208 in the Nora virus genome (reference Nora). *Red* and *blue vertical bars* represent the number of reads found at the corresponding coordinates (5′ end of sequence reads) in sense and antisense orientation, respectively. An example of coverage by nine sense reads (*red*) at position 208 and two antisense reads (*blue*) at position 226 is given at the top; 5′ ends of the sequences are indicated with *bulges*. All 186 target reads (*blue*) are used to compute the local overlap probabilities.

When *overlap_probability* is called, each position (the *query* position) of small RNAs with the appropriate size (defined by the minsize and maxsize parameters) is iteratively tested to find reads that are referenced to an overlapping position of the opposite strand, within the *scope* range. For each *query* position (*see* Fig. 2a), a local histogram of probabilities by class of overlap values is computed by normalizing the number of read (target reads) at each overlap position to 100 % over the total number of target reads up to a distance of *scope* nucleotides. In a second step, the values of local probability histograms are weighted by normalizing the number of reads at the *query* position to 100 % over the total number of reads examined in queries (Fig 2a, column "weighted probability") and then summed in a global probability histogram for the *smRNAwindow* instance (Fig. 2b, column "overlap probability"), which is eventually returned by the *overlap_probability* method. This approach ensures that each local probability histogram contributes to the final probability histogram relative to the sequencing frequency of the small RNAs for which the local probability histogram was computed.

3.3 Signature.py Output

The script returns a data frame in a tabulated text output file. In this data frame, the "overlap" column refers to the overlap values. The "pairs" column refers to the numbers of pairs found during the analysis for *all* items referenced in the input sequence dataset. In fact, the histograms returned by the *count_pairs* method for each *smRNAwindow* instance are summed. The "*z*-score" column is a standardization of the "pairs" column. It is computed by the z_score function of the script using the formula $Z_i = (P_i - \text{mean}(P))/\text{standard deviation}(P)$, where P stands for the number of pairs found for an overlap of i nucleotides. Finally, the "overlap_prob" column refers to the overall probability to find small RNA pairs overlapping with each other over the indicated overlap value. Again, note that probabilities found for each *smRNAwindow*

Fig. 2 (continued) The data frame with the local histograms for the number of pairs, the number of target reads, the overlap probabilities, and the weighted overlap probabilities are shown at the right-hand side. As 103,373 21 nt reads were matched to the Nora Virus genome in the example, the weighting factor for the query position 208 is 9/103,373. (**b**) Summed histograms of small RNA pairs, corresponding *z*-scores, and overlap probabilities for 21 nt small RNAs matched to the Nora Virus genome. (**c**) Graphic representation of the numbers of 21 nt small RNA pairs (*bottom*) and associated *z*-scores (*top*) for the indicated nucleotide overlaps, as computed in (**b**)

instances are merged in the final output, taking into account the relative weight of these instances.

Two cases of 21 nt siRNA signature analyses are given in Figs. 2 and 3. In Fig. 2, we detailed an analysis performed on Nora virus siRNAs from *Drosophila* adult thoraxes matched to a reconstituted Nora virus genome [10]. In Fig. 3, we detailed an analysis performed of endo-siRNAs immunoprecipitated with the viral protein P19 and matched to Drosophila transposon sequences ([10], and Chapter 13). In addition, we computed by hand the *z*-scores of the overlap probabilities using the formula $Zi = (p_i - \text{mean}\ (p))/\text{standard deviation}\ (p)$, where p stands for the probability of finding an overlap of i nucleotides. Note that these *z*-scores are similar but distinct from the *z*-scores found for the numbers of small RNA pairs. Indeed, the two algorithms described in this chapter can be viewed as two convergent mathematical functions whose behaviors differ at low read density and/or when the genomic distribution of the read matches varies (manuscript in preparation).

4 Notes

1. The command line to generate such a format is *bowtie -v 1 -M 1--est--trata--uppress 6,7,8 path/to/bowtie/index -f path/to/the/fasta/read/dataset* > *path/to/the/alignment/report/file.*
2. The signature.py code is built using *smRNAwindows* class instances. As the final output provides a weighted average of the values computed for each instance, values returned by instances corresponding to irrelevant items would minor values returned for relevant items. However the Class instantiation procedure used in the script allows to easily refactor python code for returning signature values for each *smRNAwindow* class instance.
3. If not made executable, the signature.py script may still be run using the syntax *python signature.py parameter-1 parameter-2 …*
4. For non-programmers, a python class may be viewed as the prototype or the template of records in a database, and an instance of this class as an actual record.
5. In addition to data ("attributes" in Object Oriented Programming), classes may contain internal functions that can transform data in the class instances or compute values from these data. These functions that are "internal" to the class instances ("private" in Objet Oriented Programming terminology) are called "methods."

a

overlap	pairs	z-score	overlap probability
1	5584	-0.322885	0.023251
2	5769	-0.248787	0.023924
3	4157	-0.894443	0.015374
4	5864	-0.210736	0.054027
5	5720	-0.268413	0.026334
6	7299	0.364025	0.030530
7	5767	-0.249588	0.023819
8	4647	-0.698183	0.016514
9	6342	-0.019283	0.032864
10	6876	0.194601	0.029577
11	4965	-0.570814	0.019876
12	6265	-0.050124	0.025330
13	5979	-0.164675	0.025141
14	5988	-0.161071	0.033796
15	6515	0.050009	0.021287
16	6578	0.075243	0.030601
17	5873	-0.207132	0.018759
18	5317	-0.429827	0.017200
19	16959	4.233154	0.122629
20	4530	-0.745045	0.017574
21	7199	0.323972	0.040412

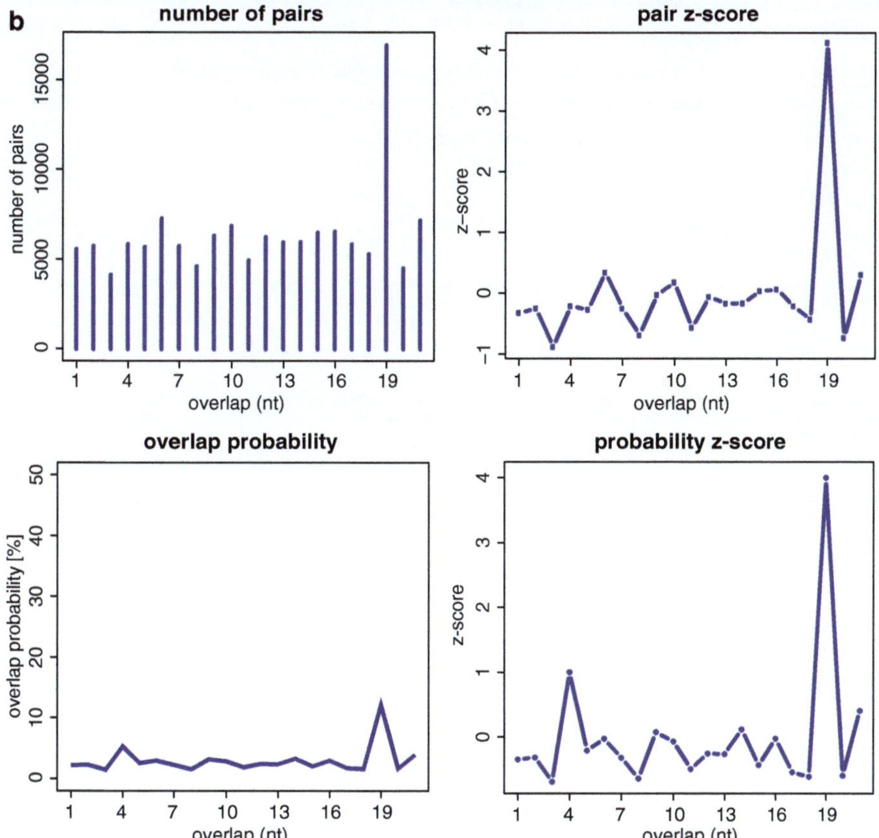

Fig. 3 Number of pairs, *z*-scores, and overlap probabilities. (**a**) An example of a data frame computed by signature.py from p19-bound siRNAs matched to Drosophila transposon sequences [9]. (**b**) Graphic representation of the number of pairs, the *z*-scores of the number of pairs (*top*), the probabilities of small RNA overlaps, and the *z*-scores of these probabilities (*bottom*) using R plot functions

Appendix: Signature.py Code

```
#!/usr/bin/env python
# computes overlap signatures from a bowtie or SAM input
# version Met Mol Biol 06-2-2012
# Usage signature.py <bowtie or sam input> <minsize> <maxsize> <minscope> <maxscope> <output>

import sys
from collections import defaultdict
from numpy import mean, std

class SmRNAwindow:
  def __init__(self, gene):
    self.readDict = defaultdict(list) # "dictionary of lists" structure {+/-offset:[size1, size2, ...], ...}
  def addread (self, polarity, offset, size):
    if polarity == "+":
      self.readDict[offset].append(size)
    else:
      self.readDict[-(offset + size -1)].append(size)
    return
  def readcount (self, lim_inf=0, lim_sup=1000):
    n=0
    for offset in self.readDict:
      licenced = [i for i in self.readDict[offset] if (i>=lim_inf and i<= lim_sup)]
      n+=len(licenced)
    return n
  def count_pairs (self, minsize, maxsize, scope):
    size_range = range (minsize, maxsize+1)
    Query_table = {}
    Target_table = {}
    frequency_table = dict ([(i, 0) for i in scope])
    for offset in self.readDict:
      for size in self.readDict[offset]:
        if size in size_range:
          Query_table[offset] = Query_table.get(offset, 0) + 1
          Target_table[offset] = Target_table.get(offset, 0) + 1
    for offset in Query_table:
      for i in scope:
        frequency_table[i] += min(Query_table[offset], Target_table.get(-offset -i +1, 0))
    for i in frequency_table:
      frequency_table[i] = frequency_table[i] / 2 # the number of smRNA PAIRS, not of the paired smRNA.
    return frequency_table
  def overlap_probability (self, minsize, maxsize, scope):
    size_range = range (minsize, maxsize+1)
    Query_table = {}
    Target_table = {}
    Total_Query_Numb = 0
    general_frequency_table = dict ([(i,0) for i in scope])
    for offset in self.readDict:
      for size in self.readDict[offset]:
        if size in size_range:
          Query_table[offset] = Query_table.get(offset, 0) + 1
          Target_table[offset] = Target_table.get(offset, 0) + 1
          Total_Query_Numb += 1
    for offset in Query_table:
      frequency_table = dict ([(i,0) for i in scope])
      number_of_targets = 0
      for i in scope:
        frequency_table[i] += Query_table[offset] *  Target_table.get(-offset -i +1, 0)
        number_of_targets += Target_table.get(-offset -i +1, 0)
      for i in scope:
        try:
          general_frequency_table[i] += (1. / number_of_targets / Total_Query_Numb) * frequency_table[i]
        except ZeroDivisionError :
          continue
    return general_frequency_table

def load_input (input_file):
  F=open(input_file)
  sampleline = F.readline()
  F.close()
  samplefields=sampleline.split()
  if len(samplefields) < 2 : #alignment format are tabulated with at least two fields.
    print "error: invalid input format"
    sys.exit()
  if samplefields[1] in ["+", "-"]: # standard tabular bowtie format detected
    F = open (input_file, "r")
    for line in F:
      fields = line.split()
      polarity = fields[1]
      gene = fields[2]
      offset = int(fields[3]) + 1 # to shift on 1-based coordinates
      size = len (fields[4])
```

```
      try:
        objDic[gene].addread (polarity, offset, size)
      except KeyError:
        objDic[gene] = SmRNAwindow(gene)
        objDic[gene].addread (polarity, offset, size)
    F.close()
  elif samplefields[0][0] == "@": # SAM format detected
    F = open (input_file, "r")
    for line in F:
      if line[0] == "@" : continue
      fields = line.split()
      if fields[2] == "*" : continue
      if fields[1] == "0": polarity = "+"
      else: polarity = "-"
      gene = fields[2]
      offset = int(fields[3])
      size = len (fields[9])
      try:
        objDic[gene].addread (polarity, offset, size)
      except KeyError:
        objDic[gene] = SmRNAwindow(gene)
        objDic[gene].addread (polarity, offset, size)
    F.close()
  else:
    print "error: invalid input format"
    sys.exit()

def z_score (table):
  value_list = [table[i] for i in sorted (table)]
  if std(value_list):
    meanlist = mean(value_list)
    stdlist = std(value_list)
    return dict ( zip ( sorted(table), [(i-meanlist)/stdlist for i in value_list] ) )
  else:
    return dict ( zip ( sorted(table), [0 for i in table]) )

objDic = {} # objDic is defined as a global variable

def __main__():
  if len(sys.argv) < 7:
    print "error: not enough parameters provided"
    sys.exit()
  load_input (sys.argv[1]) # feeds the global variable objDic (a dictionary of SmRNAwindow instances, keys=genes)
  minsize = int(sys.argv[2])
  maxsize = int(sys.argv[3])
  minscope = int(sys.argv[4])
  maxscope = int(sys.argv[5]) + 1
  general_pairs_table = dict ([(i,0) for i in range(minscope,maxscope)])
  general_prob_table = dict ([(i,0) for i in range(minscope,maxscope)])
  readcount_dic = {} # for normalized summing of local_percent_table(s)
  Total_read_in_objDic = 0
  for item in objDic:
    readcount_dic[item] = objDic[item].readcount(minsize, maxsize)
    Total_read_in_objDic += readcount_dic[item]
  OUT = open (sys.argv[-1], "w")
  for x in (objDic):
    local_pairs_table = objDic[x].count_pairs ( minsize, maxsize, range(minscope,maxscope) )
    local_prob_table = objDic[x].overlap_probability ( minsize, maxsize, range(minscope,maxscope) )
    try:
      for overlap in local_pairs_table.keys():
        general_pairs_table[overlap] = general_pairs_table.get(overlap, 0) + local_pairs_table[overlap]
    except:
      pass
    try:
      for overlap in local_prob_table.keys():
        general_prob_table[overlap] = general_prob_table.get(overlap, 0) +
(1./Total_read_in_objDic*readcount_dic[x]*local_prob_table[overlap])
    except:
      pass
  z_table = z_score(general_pairs_table)
  print >> OUT, "overlap\tpairs\tz-score\toverlap_prob"
  for overlap in sorted(general_prob_table):
    print >> OUT, "%i\t%i\t%f\t%f" % (overlap, general_pairs_table[overlap], z_table[overlap], general_prob_table[overlap])
  OUT.close()

if __name__ == "__main__" : __main__()
```

References

1. Gunawardane LS, Saito K, Nishida KM, Miyoshi K, Kawamura Y, Nagami T, Siomi H, Siomi MC (2007) A slicer-mediated mechanism for repeat-associated siRNA 5' end formation in Drosophila. Science 315:1587–1590
2. Brennecke J, Aravin AA, Stark A, Dus M, Kellis M, Sachidanandam R, Hannon GJ (2007) Discrete small RNA-generating loci as master regulators of transposon activity in Drosophila. Cell 128:1089–1103
3. Brennecke J, Malone CD, Aravin AA, Sachidanandam R, Stark A, Hannon GJ (2008) An epigenetic role for maternally inherited piRNAs in transposon silencing. Science 322: 1387–1392
4. Klattenhoff C, Xi H, Li C, Lee S, Xu J, Khurana JS, Zhang F, Schultz N, Koppetsch BS, Nowosielska A et al (2009) The Drosophila HP1 homolog Rhino is required for transposon silencing and piRNA production by dual-strand clusters. Cell 138: 1137–1149
5. Li C, Vagin VV, Lee S, Xu J, Ma S, Xi H, Seitz H, Horwich MD, Syrzycka M, Honda BM et al (2009) Collapse of germline piRNAs in the absence of Argonaute3 reveals somatic piRNAs in flies. Cell 137:509–521
6. Malone CD, Brennecke J, Dus M, Stark A, McCombie WR, Sachidanandam R, Hannon GJ (2009) Specialized piRNA pathways act in

germline and somatic tissues of the Drosophila ovary. Cell 137:522–535

7. Lau NC, Robine N, Martin R, Chung WJ, Niki Y, Berezikov E, Lai EC (2009) Abundant primary piRNAs, endo-siRNAs, and microRNAs in a Drosophila ovary cell line. Genome Res 19:1776–1785
8. de Vanssay A, Bouge AL, Boivin A, Hermant C, Teysset L, Delmarre V, Antoniewski C, Ronsseray S (2012) Paramutation in Drosophila linked to emergence of a piRNA-producing locus. Nature 490:112–115
9. Fagegaltier D, Bouge AL, Berry B, Poisot E, Sismeiro O, Coppee JY, Theodore L, Voinnet O, Antoniewski C (2009) The endogenous siRNA pathway is involved in heterochromatin formation in Drosophila. Proc Natl Acad Sci U S A 106:21258–21263
10. van Mierlo JT, Bronkhorst AW, Overheul GJ, Sadanandan SA, Ekstrom JO, Heestermans M, Hultmark D, Antoniewski C, van Rij RP (2012) Convergent evolution of argonaute-2 slicer antagonism in two distinct insect RNA viruses. PLoS Pathog 8:e1002872
11. Langmead B, Trapnell C, Pop M, Salzberg SL (2009) Ultrafast and memory-efficient alignment of short DNA sequences to the human genome. Genome Biol 10:R25

Chapter 13

Isolation of Small Interfering RNAs Using Viral Suppressors of RNA Interference

Marius van den Beek, Christophe Antoniewski, and Clément Carré

Abstract

The *tombusvirus* P19 VSR (*v*iral *s*uppressor of *R*NA interference) binds siRNAs with high affinity, whereas the *Flockhouse Virus* (FHV) B2 VSR binds both long double-stranded RNA (dsRNA) and small interfering RNAs (siRNAs). Both VSRs are small proteins and function in plant and animal cells. Fusing a *N*uclear *L*ocalization *S*ignal (NLS) to the N-terminus shifts the localization of the VSR from cytoplasmic to nuclear, allowing researchers to specifically probe the subcellular distribution of siRNAs, and to investigate the function of nuclear and cytoplasmic siRNAs. This chapter provides a detailed protocol for the immunoprecipitation of siRNAs bound to epitope-tagged VSR and subsequent analysis by 3′-end-labeling using cytidine-3′,5′-*bis* phosphate ([5′-^{32}P]pCp) and northern blotting.

Key words RNAi, siRNA, Endo-siRNA, Immunoprecipitation, Viral suppressor of RNAi, B2, P19, RNA purification, pCp labeling, RNA detection, *Drosophila*

1 Introduction

Small interfering RNAs (siRNAs) are implicated in a variety of processes such as *T*ransposable *E*lements (TE) repression [1], maintenance of pericentric heterochromatin [2], and antiviral defense [3]. In *Drosophila,* the biogenesis of siRNAs starts with the cleavage of a long double-stranded RNA (dsRNA) precursor by the Dicer-2 endonuclease into a 21 nucleotide RNA duplex structure with 2 nt 3′OH overhangs [4]. Following cleavage, the siRNA duplex is loaded into an Argonaute (Ago) containing RNA-induced silencing complex, where the duplex is unwound and the strand bearing the thermodynamically more stable 5′ end (passenger strand) is cleaved by the central RISC protein Argonaute (Ago2 in flies) [5, 6]. The remaining strand (guide strand) is 2′-O-methylated by Hen1 at its 3′ end and guides the recognition and subsequent cleavage of complementary single-stranded target RNAs [7].

Andreas Werner (ed.), *Animal Endo-siRNAs: Methods and Protocols*, Methods in Molecular Biology, vol. 1173, DOI 10.1007/978-1-4939-0931-5_13, © Springer Science+Business Media New York 2014

As antiviral RNAi limits viral replication, many viruses evolved suppressors of RNAi (VSRs) [8]. Depending on the VSR, RNAi suppression may occur through the binding of long dsRNA substrates and/or siRNA duplexes (B2, P19, DCV1A), thereby limiting substrate availability for Dicer and Argonautes, or through the direct inhibition of Argonaute proteins (VSR-1A of Cricket Paralysis Virus) [9]. In general, VSRs that act through binding to dsRNA or siRNAs are active in both plants and animals.

The *tombusvirus* P19 VSR is a 19 kD protein that forms a head-to-tail homodimer and localizes to the cytoplasm when expressed as a transgene in *Drosophila* S2 cells and salivary glands [2]. P19 binds 21 nucleotide dsRNAs [10, 11], thereby suppressing siRNA-mediated posttranscriptional gene silencing (PTGS) in plants [12], insects [13], and mammalian cells [14, 15], leading to a derepression of endogenous siRNA (endo-siRNA) targeted transposable elements. We have shown that fusing P19 to the NLS of the transformer (tra) protein efficiently re-localizes it to the nucleus [2].

FHV B2 is 12kD in size and forms a four-helix bundle that binds to one face of an A-form RNA duplex, independent of its length, thereby both limiting the processing of long dsRNA by Dicer and siRNA duplex incorporation into RISC [16]. Additionally, B2 has been reported to bind to the conserved PAZ domain of Dicer family proteins [17]. B2 immunoprecipitates efficiently long dsRNA, but not siRNAs, likely due to the inhibition of siRNA biogenesis. When expressed in *Drosophila* salivary glands B2 localizes to nucleoli, the nucleoplasm, and the cytoplasm [2].

Through the immunoprecipitation of a nuclear-targeted P19 we were able to pull down siRNAs that localize to the nucleus without prior biochemical fractionation and to compare them to siRNAs that were pulled down by immunoprecipitating cytoplasmic P19 or the dsRNAs that co-precipitate with B2 [2]. RNAi against mRNAs is not suppressed by the NLS-P19 transgene, indicating that the bulk of siRNA-mediated PTGS is not occurring inside the nucleus. Instead, a redistribution of H3K9me2 and Heterochromatin Protein-1 (HP1) is observed when expressing NLS-P19 but not unmodified P19. This mirrors the redistribution of repressive chromatin marks observed in mutants of the RNAi pathway and links endo-siRNAs to the maintenance of chromatin organization [2]. This also highlights the potential for nuclear-engineered VSRs in dissecting the contribution of RNAi to heterochromatin maintenance.

Here we describe the use of the V5-tagged VSRs P19, NLS-P19, and B2 to immunoprecipitate bound siRNAs for experiments such as 3′-end-labeling using [5′-^{32}P]pCp, northern blot detection, or high-throughput sequencing (*see* Antoniewski C, Chapter 12).

2 Materials

2.1 Immunoprecipitation

1. Mouse Monoclonal anti-V5 (Invitrogen, cat. no R960-25) and/or rat anti-HA High Affinity antibodies (Roche, cat. no 11867423001).
2. Gammabind-G sepharose (GE Healthcare).
3. Phosphate Buffered saline (PBS) (10× stock): 1.37 M NaCl, 27 mM KCl, 100 mM Na_2HPO_4. Adjust to pH 7.4 with HCl; autoclave. Store at room temperature.
4. Lysis buffer: 50 mM Tris–HCl pH 7.5, 150 mM NaCl, 2.5 mM $MgCl_2$, 250 mM sucrose, 0.05 % Nonidet P-40, 0.5 % Triton X-100. Before use, adjust to 1 mM Dithiothreitol (DTT), 1× protease inhibitor mixture cocktail (Roche). Store at 4 °C. Before use add 40 U per ml of RNase inhibitor.
5. RNase inhibitor: RNase OUT (Invitrogen).
6. Wash buffer: 50 mM Tris–HCl pH 7.5, 150 mM NaCl, 2.5 mM $MgCl_2$, 250 mM sucrose, 0.05 % Nonidet P-40, 0.5 % Triton X-100. Store at 4 °C. Before use add 1 mM Dithiothreitol (DTT), 1× protease inhibitor mixture cocktail and 40 U per ml of RNase OUT.
7. RNA Loading Dye (2×) (New England Biolabs). Store at −20 °C.
8. 2× Laemmli Buffer: 4 % SDS, 20 % glycerol, 10 % 2-mercaptoethanol, 0.004 % bromophenol blue, 0.125 M Tris–HCl pH 6.8. Store at −20 °C.
9. 100 mM $CuSO_4$ solution in distilled water. Store at 4 °C.
10. Rotating wheel.

2.2 RNA Extraction and pCp Labeling

1. Nuclease free water.
2. Isopropanol.
3. Chloroform.
4. Phenol/Chloroform/Isoamyl alcohol pH 4.5.
5. 80 % Ethanol (in RNase-free water).
6. TRIzol reagent (Sigma).
7. RNA carrier (glycogen or homemade linear acrylamide).
8. 3 M NaAcetate (NaAc), pH 5.2.
9. T4 RNA ligase (Roche).
10. pCp (Cytidine 5′-triphosphate disodium salt), [5′-^{32}P]-3000 Ci/mmol 10 mCi/ml (PerkinElmer).
11. Dimethyl sulfoxide (DMSO).
12. G50 MicroSpin columns (GE Healthcare).
13. 15 % acrylamide denaturing gel: 1× TBE buffer (89 mM Tris–borate, 2 mM EDTA), 15 % acrylamide/bisacrylamide (19:1), 7 M urea, TEMED, and 10 % APS fresh solution.

3 Methods

The use of epitope-tagged P19, NLS-P19, and B2 to immunoprecipitate small RNAs was described by us and others (*see* Subheading 1). Depending on the amount required for the small RNA analysis one might use transient transfection or establish a stable cell line. To analyze small RNAs by northern blotting or $^{32P\text{-}pCp}$ 3′ end-labeling, transient transfection is sufficient and takes less time. To immunoprecipitate large amounts of small RNAs for deep-sequencing or to detect low-abundant RNAs, stable transfections are more suitable but require more time. We provide instructions for the immunoprecipitation from transient and stable transfections (*see* **Note 1**). Entry clones and expression vectors may be obtained from the authors and were described previously by Fagegaltier et al. [2].

3.1 Immunoprecipitation of Small RNAs Bound to Epitope-Tagged VSR in Drosophila S2 Cells

3.1.1 Using Transient Expression of the VSRs

1. Transfect 4 μg of pMT-DEST48 control plasmid, pMT-B2-V5, pMT-P19-V5, or pMT-NLSP19-V5 in $3*10^6$ S2 cells with Effectene Reagent (Invitrogen) according to the manufacturer's instructions.
2. Depending on the number of conditions tested, cells can be split 2 days after transfection and cultured for an additional 2 days.
3. Induce construct expression with 500 μM $CuSO_4$ for 24 h.
4. Equilibrate 70 μL of beads for 10 min in lysis buffer at 4 °C on a rotating wheel, 15 rpm.
5. Harvest cells, wash twice in cold 1× PBS, and lyse on ice for 30 min in 1 ml of lysis buffer.
6. Centrifuge at 16,100 ×*g* for 15 min at 4 °C to pellet insoluble cell debris.
7. Transfer 10 % of supernatant to a new microcentrifuge tube. This will be the input sample for the analysis of the IPs.
8. Pre-clear the extract: Transfer the remaining supernatant to a new tube and add 20 μL of equilibrated beads (from **step 4**). Incubate for 1 h at 4 °C on a rotating wheel, 15 rpm.
9. Centrifuge the pre-clearing mix for 5 min at 4 °C, 800 rpm and transfer the cleared supernatant to a new microcentrifuge tube.
10. Add 5 μg of mouse anti-V5 antibody (*see* **Note 2**) and 50 μL of fresh equilibrated Gammabind-Plus resin slurry (*from* **step 4**) to the pre-cleared supernatant and incubate overnight at 4 °C on a rotating wheel (15 rpm).
11. Centrifuge the samples at 4 °C and 800 rpm for 5 min.
12. Keep the supernatant as the unbound fraction.
13. Wash the beads five times in 1 ml of wash buffer (*see* **Note 3**). *Continue with Subheading 3.1.3.*

3.1.2 Immunoprecipitation from Stable Cell Line Expressing a VSR

It is also possible to establish blasticidin or hygromycin-resistant S2 cell lines stably transformed with the appropriate vector (in our case pAWH-P19 or pAWHNLS-P19 constructs) using Effectene Reagent (Invitrogen) according to the manufacturer's instructions (*see* http://www.flyrnai.org/DRSC-PRL.html for specific S2 stable cell line establishment).

1. Harvest cells from ten to fifteen 75 cm^2 plates at 80 % confluency (*see* **Note 4**), wash twice in cold PBS 1×, and lyse the cells on ice for 30 min in lysis buffer.
2. Centrifuge at 14,000 rpm for 15 min at 4 °C to pellet insoluble cell debris.
3. Transfer 10 % of supernatant to a new microcentrifuge tube. This will be the input sample for the analysis of the IPs.
4. Pre-clear the extract: Transfer the remaining supernatant to a new tube and add 50 μL of equilibrated beads for 1 h at 4 °C on a rotating wheel, 15 rpm (*see* **Note 5**).
5. After centrifugation at 800 rpm for 5 min at 4 °C add 400 μL of equilibrated Gammabind-Plus resin (see **step 4** of Subheading 3.1.1) and 20 μg of the appropriate antibody (here: rat anti-HA High Affinity) to the pre-cleared supernatant and incubate for 2 h at 4 °C on a rotating wheel at 15 rpm.
6. After centrifugation at 800 rpm for 5 min at 4 °C wash beads five times in wash buffer.

3.1.3 Purification of Immunoprecipitated RNAs

1. Dilute 25 % of the beads in 1× Laemmli buffer for protein analysis by western blotting. Store the sample at −20 °C.
2. Wash the remaining beads in wash buffer without proteinase inhibitors.
3. Incubate beads with 20 μL of proteinase K for 2 h (20 μg with an activity of 30 U/mg).
4. Add 400 μL of TRIzol and 100 μL of chloroform directly to the beads.
5. Vortex the mix for 15 s.

   (at this step, sample can be stored at −80 °C if needed).
6. Incubate sample at room temperature for 3 min.
7. Centrifuge at 14,000 rpm at 4 °C for 15 min.
8. Carefully pipette aqueous phase (upper phase) into a clean microcentrifuge tube and discard the lower phase.
9. Add an equal volume (around 140 μL) of isopropanol and mix by gentle inversion (1 μL of glycogen (20 μg/μl) or linear acrylamide (5 μg) carrier facilitates precipitation and visualization of the RNA pellet).
10. Incubate sample at room temperature for 10 min.
11. Centrifuge tubes at 14,000 rpm at 4 °C for 15 min.

12. Discard the supernatant without touching the pellet and wash with 80 % ethanol, vortex briefly to detach the pellet from the tube.
13. Centrifuge at 14,000 rpm at 4 °C for 10 min.
14. Carefully remove the ethanol and air-dry the pellet for about 5–8 min (a visible white pellet should disappear during the drying).
15. Add 30 μL of RNase-free water to the pellet and resuspend by gently pipetting up and down.

RNA concentration and quality can be checked using a nanodrop spectrophotometer (expected concentration for IPs using the indicated amount of cells and proteins is around 10 (transient expression protocol) to 100 ng/μL (stable expression protocol)). The RNA can be used immediately or stored at −80 °C (avoid repeated freezing/unfreezing cycles).

For total RNA analysis (input or supernatant after IPs), RNA should be extracted with TRIzol Reagent (Invitrogen) according to the manufacturer's instructions, except that RNA washes are performed in 80 % ethanol.

At this point, RNA from the IPs can be used for standard RNA analysis (RT-qPCR or northern blotting for example). However, in this chapter we focus on [5′-^{32}P]pCp labeling followed by northern blotting.

3.1.4 32P-pCp 3′ End-Labeling RNA

RNA molecules can be 3′-end-labeled using [5′-^{32}P]pCp (cytidine-3′,5′-*bis* phosphate) and RNA ligase. The RNA to be labeled must have a free 3′-hydroxyl end for better results. It was shown however that pCp labeling is efficient enough to label small RNA (piRNA) although they are 2′-O-methylated at their 3′-OH end [18]. Using this protocol, we aimed at detecting endo-siRNAs IP with VSRs proteins. This class of small RNA is 2′O-methylated at their 3′-extremity after their passage into the Ago2–RISC complex. The 2′O-methylation at the 3′-OH end could affect the efficiency of the pCp 3′-end labeling due to the inaccessibility of the 2′ end of small RNAs. However and importantly, the VSRs used to precipitate the siRNAs in this protocol capture them as a duplex before entry into the Ago2–RISC complex and subsequent 2′O-methylation.

1. Pipette 4 μL of RNA in a new microcentrifuge tube. This corresponds to around 10 % of the immunoprecipitated RNA. More RNA may be labeled if the signal obtains is too low.
2. Add 2.5 μL, 100 μCi of [5′-^{32}P]pCp.
3. Add 3 μL 10× RNA ligase buffer.
4. Add 3 μL DMSO (*see* **Note 6**).
5. Add water to 29.5 μL total volume.
6. Add 0.5 μL T4 RNA ligase (10 U).

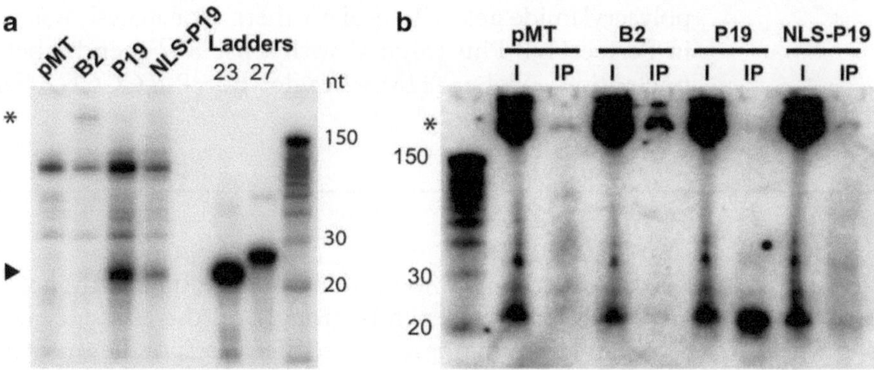

Fig. 1 P19 and B2 respectively sequester endogenous TE-matching siRNAs or longer precursors in S2 cells. (**a**) Immunoprecipitated P19 and NLS-P19 sequester 21 nt RNAs that migrate as 22–23 nt species after 3′end-pCp labeling (*arrowhead*) whereas larger RNA species are sequestered exclusively by B2 (*asterisk*). Control immunoprecipitation (pMT) (*see* **Note 7**) was performed using S2 cells transfected with the empty expression vector pMT-DEST48 (from Fagegaltier et al., PNAS, 2009). (**b**) A sense HMS-Beagle probe revealed enriched endo-siRNAs in Northern blots of P19 and NLS-P19 RNAs immunoprecipitates (IP) and longer RNA species (*asterisk*) in B2 RNA immunoprecipitate. I corresponds to total RNA input material (from Fagegaltier et al., PNAS) [2]

7. The 30 μL reactions are then incubated overnight at 4 °C (*see* **Note 8**).

   At this step, samples can be stored several days at −20 °C.
8. Add 100 μL of H_2O to the pCp labeled samples.
9. Remove unincorporated nucleotides using a G50-column (GE Healthcare).
10. Pipette 40 μL of 3 M NaAc, 260 μL H_2O, and 2 μL of glycogen to the samples.
11. Add 400 μL of Phenol/Chloroform/Isoamyl alcohol and proceed to RNA classical extraction. Carefully pipette aqueous phase (upper phase) into a clean Eppendorf tube and precipitate with 1 ml of ethanol for 2 h at −20 °C. Centrifuge 30 min at 4 °C, 14,000 rpm. Wash with 70 % ethanol and let the tubes dry at room temperature (*see* **Note 9**).
12. Add 12 μL of loading RNA buffer 1×.
13. Denature the samples for 4 min at 95 °C and load them onto a 7 M urea denaturing 15 % polyacrylamide gel.
14. Signals are visualized by autoradiography (*see* Fig. 1a and **Notes 6** and **10**).

Northern blot analysis of IP RNAs was able to confirm the presence of endo-siRNA from the HMS-Beagle retrotransposon (Fig. 1b). Briefly, 20 μg of total RNAs isolated from transfected cells (Input), or 90 % of the immunoprecipitated RNAs (IP) were resolved by electrophoresis onto 7 M urea denaturing 15 %

polyacrylamide gels. Classical northern blot analysis was performed in PerfectHyb Plus (Sigma) with sense 5′-^{32}P end-labeled oligonucleotide probe: HMS-Beagle 5′-^{32}P-*TCCCGACATTCCA TAGGCATTTA*-3′.

4 Notes

1. IP experiments were done successfully with less material. However, this depends strongly on the efficiency of the antibody to immunoprecipitate the corresponding protein.
2. Here, monoclonal anti-V5 antibody (Invitrogen) and anti-HA (Roche) were used. However, others appropriate tag antibodies such anti-Flag as well as specific antibodies against VSRs can be used.
3. Salt concentration in the wash buffer is 150 mM NaCl. If background problems occur salt concentration can be raised up to 800 mM KCl.
4. The number of 75 cm^2 plate used from stable cell line depends on the expression of each individual stable cell line. In our hands, RNA IP of 15 plates was always sufficient for a good pCp labeling reaction using the VSRs described in this chapter.
5. Pre-clearing can be done overnight to decrease background due to nonspecific protein binding to the beads.
6. DMSO seems to improve end-labeling especially with difficult to label RNAs. However, higher concentration of DMSO considerably inhibits ligase activity.
7. Empty vector control is absolutely recommended to detect unspecific or artifactual signal of pCp reaction (see pMT line in Figure. A).
8. For the pCp reaction, 4 °C is the recommended temperature. However, reaction can be done at 37 °C during 4 h if needed.
9. The pCp reaction as described above should give several million *cpm* of labeled RNA.
10. Small RNAs immunoprecipitated from S2 cells (control) and stably transformed P19 and NLS-P19 S2 cells were previously cloned for sequencing using the DGE-Small RNA Sample Prep Kit and the Small RNA Sample Prep v1.5 Conversion Kit from Illumina, following manufacturer instructions (*see* ref. [2]).

    Outline of the methods described in this chapter:

    - Transfection (transitory) and induction of VSR constructs (4–5 days).
    - Isolation of co-immunoprecipitated RNA (1 day).
    - Analysis by [5′-^{32}P]pCp 3′ end labeling and/or Northern Blot (2 days).

Acknowledgments

We thank D. Kirschner and B. Berry for VSR expression vectors and D. Fagegaltier and A.L. Bougé for fruitful discussions. This work was supported by postdoctoral fellowships from the *Agence Nationale de la Recherche* to C.C. (grant number ANR BLAN 1210 01 "Nuclear endosiRNAs" to C.A.) and Ph.D. fellowships from the French government to M.vd.B.

References

1. Siomi MC, Saito K, Siomi H (2008) How selfish retrotransposons are silenced in Drosophila germline and somatic cells. FEBS Lett 582:2473–2478
2. Fagegaltier D, Bouge AL, Berry B, Poisot E, Sismeiro O, Coppee JY, Theodore L, Voinnet O, Antoniewski C (2009) The endogenous siRNA pathway is involved in heterochromatin formation in Drosophila. Proc Natl Acad Sci U S A 106:21258–21263
3. Ding SW, Voinnet O (2007) Antiviral immunity directed by small RNAs. Cell 130:413–426
4. Hammond SM (2005) Dicing and slicing: the core machinery of the RNA interference pathway. FEBS Lett 579:5822–5829
5. Khvorova A, Reynolds A, Jayasena SD (2003) Functional siRNAs and miRNAs exhibit strand bias. Cell 115:209–216
6. Schwarz DS, Hutvagner G, Du T, Xu Z, Aronin N, Zamore PD (2003) Asymmetry in the assembly of the RNAi enzyme complex. Cell 115:199–208
7. Horwich MD, Li C, Matranga C, Vagin V, Farley G, Wang P, Zamore PD (2007) The Drosophila RNA methyltransferase, DmHen1, modifies germline piRNAs and single-stranded siRNAs in RISC. Curr Biol 17:1265–1272
8. Li F, Ding SW (2006) Virus counterdefense: diverse strategies for evading the RNA-silencing immunity. Annu Rev Microbiol 60:503–531
9. Nayak A, Berry B, Tassetto M, Kunitomi M, Acevedo A, Deng C, Krutchinsky A, Gross J, Antoniewski C, Andino R (2010) Cricket paralysis virus antagonizes Argonaute 2 to modulate antiviral defense in Drosophila. Nat Struct Mol Biol 17:547–554
10. Rawlings RA, Krishnan V, Walter NG (2011) Viral RNAi suppressor reversibly binds siRNA to outcompete Dicer and RISC via multiple turnover. J Mol Biol 408:262–276
11. Vargason JM, Szittya G, Burgyan J, Hall TM (2003) Size selective recognition of siRNA by an RNA silencing suppressor. Cell 115: 799–811
12. Silhavy D, Molnar A, Lucioli A, Szittya G, Hornyik C, Tavazza M, Burgyan J (2002) A viral protein suppresses RNA silencing and binds silencing-generated, 21- to 25-nucleotide double-stranded RNAs. EMBO J 21:3070–3080
13. Lakatos L, Szittya G, Silhavy D, Burgyan J (2004) Molecular mechanism of RNA silencing suppression mediated by p19 protein of tombusviruses. EMBO J 23:876–884
14. Dunoyer P, Lecellier CH, Parizotto EA, Himber C, Voinnet O (2004) Probing the microRNA and small interfering RNA pathways with virus-encoded suppressors of RNA silencing. Plant Cell 16:1235–1250
15. Liu X, Houzet L, Jeang KT (2012) Tombusvirus P19 RNA silencing suppressor (RSS) activity in mammalian cells correlates with charged amino acids that contribute to direct RNA-binding. Cell Biosci 2:41
16. Chao JA, Lee JH, Chapados BR, Debler EW, Schneemann A, Williamson JR (2005) Dual modes of RNA-silencing suppression by Flock House virus protein B2. Nat Struct Mol Biol 12:952–957
17. Singh G, Popli S, Hari Y, Malhotra P, Mukherjee S, Bhatnagar RK (2009) Suppression of RNA silencing by Flock house virus B2 protein is mediated through its interaction with the PAZ domain of Dicer. FASEB J 23:1845–1857
18. Saito K, Nishida KM, Mori T, Kawamura Y, Miyoshi K, Nagami T, Siomi H, Siomi MC (2006) Specific association of Piwi with rasiRNAs derived from retrotransposon and heterochromatic regions in the Drosophila genome. Genes Dev 20(16):2214–2222

Chapter 14

Computational Analysis, Biochemical Purification, and Detection of tRNA-Derived Small RNA Fragments

Simon P. Keam, Andrew Sobala, David T. Humphreys, Catherine M. Suter, and Gyorgy Hutvagner

Abstract

The rapidly growing list of small RNA species generated by next-generation sequencing technologies has accelerated the development of new bioinformatics tools for their detection. Small RNAs generated from tRNAs, transfer RNA-derived fragments (tRFs), represent a novel challenge in accurately identifying and distinguishing them from random degradation products of tRNAs. Here, we describe a bioinformatics approach to detect tRFs in next-generation sequencing libraries. We also present a biochemical purification protocol for enriching 5′ tRFs and separating them from miRNAs. And finally, we suggest reliable methods for detecting and quantifying tRFs.

Key words Transfer RNA fragments, tRF, Deep sequencing, Bioinformatics, Ionic exchange chromatography, Northern blotting, Quantitative PCR

1 Introduction

Current methods for the detection and quantification of short RNAs have been primarily driven by the abundant and well-characterized miRNAs. This has resulted in the development of sensitive northern blotting [1] and various PCR amplification strategies specifically optimized for their detection. However, the advent of deep sequencing technology has unearthed entirely novel classes of small RNAs with novel features and functions. One such class of small RNAs is those derived from transfer RNAs (tRNAs) [2], commonly referred to as tRFs [3]. tRFs are processed from highly abundant precursors, and this introduces challenges in distinguishing them from the RNA degradome in high-throughput sequencing data. tRFs are processed from the 5′ and 3′ end of the mature or primary tRNAs and some of these have been shown to possess the

These two authors are equal contributors.

Andreas Werner (ed.), *Animal Endo-siRNAs: Methods and Protocols*, Methods in Molecular Biology, vol. 1173,
DOI 10.1007/978-1-4939-0931-5_14, © Springer Science+Business Media New York 2014

ability to repress global translation [3–6] or specifically silence mRNA targets that have a complementary sequence [7, 8]. Here, we describe two methods, one bioinformatical and one biochemical, to assist with the analysis of tRFs. Specifically, the bioinformatics method explains the key steps required for the detection and quantification of tRFs in raw deep sequencing data, which is not a typical component of modern mapping pipelines. This involves using an alternate mapping strategy based on publically available tRNA databases and subsequent bioinformatic processes. The biochemical method describes a purification protocol that enriches tRFs and separates them from major cellular small RNA components such as miRNAs and siRNAs. This method involves successive rounds of anionic exchange, gel filtration, and polyacrylamide gel (PAGE) purification.

2 Materials

2.1 Reagents

1. Hypotonic lysis buffer: 10 mM HEPES pH 7.9, 10 mM KCl, 1.5 mM $MgCl_2$, 0.5 mM DTT.
2. Buffer B: 300 mM HEPES pH 7.9, 1.4 M KCl, 30 mM $MgCl_2$.
3. Buffer D: 20 mM HEPES pH 7.9, 20 % (v/v) glycerol, 100 mM KCl, 0.2 mM EDTA, 9.5 mM DTT.
4. Ionic Buffer A: 20 mM Tris–Cl pH 8.0, 10 mM KCl, 1.5 mM $MgCl_2$, 0.5 mM DTT.
5. Ionic Buffer B: 20 mM Tris–Cl pH 8.0, 0.5 M NaCl, 10 mM KCl, 1.5 mM $MgCl_2$, 0.5 mM DTT.
6. Buffer E: 20 mM Tris–Cl pH 7.5, 150 mM NaCl, and 1 mM DTT.
7. PK buffer: 200 mM Tris–Cl pH 7.5, 25 mM EDTA, 300 mM NaCl, 2 % SDS (v/v), and 200 μg/mL proteinase K (Ambion).
8. Gel filtration standards (Bio-Rad).
9. Cells (HeLa) or pre-prepared cytosolic S100 cell extract.
10. Formamide Loading Buffer: 2.5 mM EDTA, 49 % (v/v) deionized formamide, 0.1 % (w/v) bromophenol blue.
11. Reagents for denaturing urea-PAGE gel: acrylamide, urea, TEMED, ammonium persulfate.
12. 10 bp radiolabeled Decade RNA ladder (Applied Biosystems).
13. Glycogen (Ambion).
14. 70 % (v/v) ethanol.
15. 100 % (v/v) ethanol.

2.2 Equipment

1. HPLC apparatus.
2. Anionic exchange column (Q column).
(e.g., HiTrap Q HP 1 mL column (GE Healthcare).

3. Gel filtration chromatography column,
   e.g., Superdex-200 XK16/60 gel filtration column (GE Healthcare).
4. Ultracentrifuge.
5. Swinging bucket rotor capable of 60,000 × g with appropriate tubes (e.g., Beckman SW-41 Ti).
6. Dialysis equipment.
7. Dounce homogenizer.
8. PAGE gel running apparatus.

2.3 Websites and tRNA Databases

1. The UCSC genome browser.
   This site is a publicly available database that allows the uploading and visualization of deep sequencing data in the form of coverage tracks [9].
   http.://genome.ucsc.edu.
2. The Genomic tRNA database (gtRNAdb).
   This is a free database that contains predicted tRNA genes generated by the tRNAscan-SE algorithm from Eukaryotic and Prokaryotic genome data [10].
   http://gtrnadb.ucsc.edu/.
   Specific information for tRNAs on gtRNAdb (human only):
   http://gtrnadb.ucsc.edu/Hsapi19/Hsapi19-summary.html.
3. The Lowe Lab tRNAscan-SE Search Server.
   This site allows you to search for the similarity of input RNA sequences to known tRNAs and is the basis for the tRNAs found in the genomic tRNA database [11].
   http://lowelab.ucsc.edu/tRNAscan-SE.
4. miRBase.
   This is the primary online source of microRNA sequences and annotations [12].
   ftp://mirbase.org/pub/mirbase/CURRENT/hairpin.fa.gz.

3 Methods

3.1 Mapping to a Custom tRNA Reference

This section describes using the tRNA sequences listed in gtRNAdb to specifically map deep sequencing data to identify tRFs. This is done to circumvent problems that can arise when repetitive sequences (such as tRFs) map more than once to the reference genome. These reads mapping to multiple sites can be inadvertently discarded by standard mapping pipelines that do not permit multiple mapping. In addition, using this method also avoids the need to generate complex transcriptome references. This section outlines a strategy to map and interpret tRF data in deep sequencing libraries (*see* **Note 1**).

1. Download the tRNA database from gtRNAdb. This can be found at:

   http://gtrnadb.ucsc.edu/download.html.

2. Next, filter this dataset to select only tRNAs for your model organism, be careful to maintain the fasta sequence format. The database download does contain a significant number of duplicate records and these should be removed. For example there are a total of 1,256 *Homo sapiens* entries, but only 631 are unique (*see* **Note 2**).
3. Map the raw sequencing data to the compiled tRNA database made in **step 2**. It is recommended to not allow any mismatches or multi-mappers as tRNAs encode little inherent variation (*see* **Note 3**). This and subsequent alignments will produce Sequence Alignment/Map (SAM) files which can be analyzed to assess and quantify the abundance of fragments. All unaligned reads should be saved to a separate file for subsequent mapping as described in **step 4**.
4. The unaligned reads from **step 3** will most likely contain a significant proportion of sequences that can be aligned to other regions of the genome, particularly miRNAs. It is important to identify tags that represent genuine transcripts as these data are used for normalization. To maximize the proportion of mapped tags we suggest mapping the unaligned reads from **step 2** to miRNA hairpin sequences of your model organism (*see* **Note 4**). Reads that align to miRBase should be counted, while those that do not are kept to be aligned to the full reference genome. At the end of this step there should be three SAM/BAM files (one each for tRNA, miRNA, and genome).

3.2 Organization and Normalization of Reads Derived from tRNAs

Determining differences in the overall abundance of tRFs, and particularly those from different tRNA isotypes, is essential to determining changes under different cellular conditions. This section describes the basics of organizing and normalizing the reads to infer meaningful conclusions which may be confirmed by conventional techniques such as northern hybridization or PCR.

1. Using the tRNA SAM file generated in Subheading 3.1, collate the counts for each tRNA entry according to the gtRNAdb identifiers in the downloaded fasta dataset.
2. Cross-reference the gtRNAdb identifier in the SAM file with the downloaded fasta information for all gtRNAdb entries, as shown in Fig. 1. This can be used to organize tRFs by many different features (*see* **Note 5**) including:
   - Chromosome of origin.
   - tRNA isotype.
   - Anticodon.

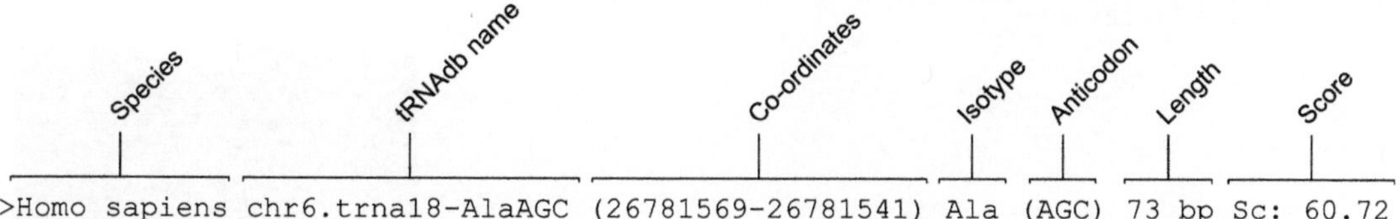


```
>Homo_sapiens_chr6.trna18-AlaAGC (26781569-26781541) Ala (AGC) 73 bp Sc: 60.72
GGGGAATTAGCTCAAGTGGTAGAGCGCTTGCTTAGCATGCAAGAGGTAGTGGGATCAATGCCCACATTCTCCA
```

Fig. 1 gtRNAdb information for human tRNA locus chr6.trna18-AlaAGC in the Genomic tRNA Database fasta format

- Chromosome position.
- Score (*see* **Note 6**).
- Source tRNA length.

3. To accurately compare tRF counts between different datasets/libraries each dataset must be normalized. We recommend normalizing reads as a fraction of total mapped reads, i.e., for each tRNA isoform calculate the number of sequences that align to it and divide it by the total number of reads that map to the sum of tRNA, miRNA, and genome mapped reads. This ensures that results are not biased by normal variations in the depth of sequencing.
4. If differences in the distribution of tRFs from different isotypes and/or anticodons are of interest, normalize these subcategories to the total number of tRFs (*see* **Note 7**).

3.3 Generation of WIG Tracks for Visualization on Genome Browser

Since tRFs are shorter than the source tRNA, the visualization of tRFs in relation to the source tRNA can indicate the part of the molecule that they derive from. Accordingly, tRFs can be subdivided into three main categories; 5′ tRFs, 3′ U tRFs, and 3′ CCA tRFs [13]. The Wiggle track format (WIG) indicates how many times each nucleotide is identified in the deep sequencing dataset. Data in this format can be loaded into genome browsers to visualize the coverage of each sequenced nucleotide. Here are the steps required to achieve this:

1. Convert your mapped deep sequencing data into wiggle track files. Detailed information for this procedure can be found at:

   https://cgwb.nci.nih.gov/goldenPath/help/wiggle.html.
2. Upload the correctly formatted WIG tracks to the UCSC genome browser:

   http://genome.ucsc.edu/cgi-bin/hgGateway (*see* **Note 8**).
3. Navigate to the tRNA loci of choice. Toggle the reference tRNA gene tracks by navigating to "Genes and Gene Prediction tracks" and selecting the "tRNA genes" track to "Full" and press "Refresh." The final browser window should appear similar to Fig. 2 (*see* **Notes 9** and **10**). This view can predict the sites where the putative tRF is most likely to be processed.

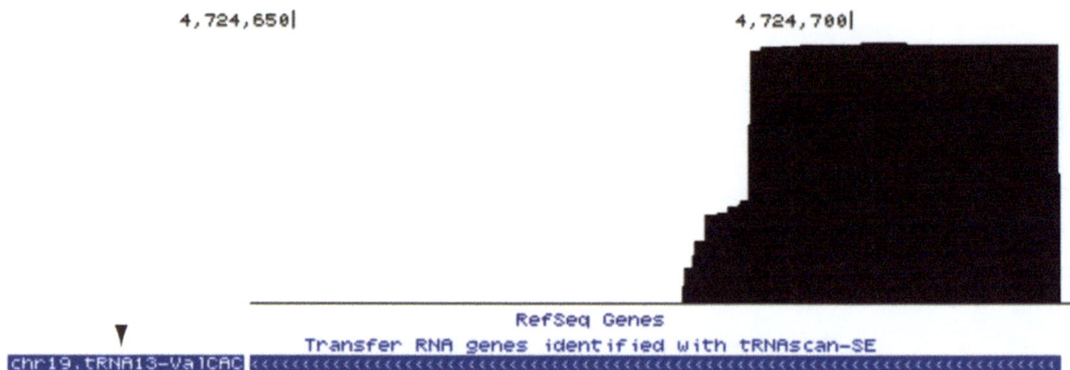

Fig. 2 Browser information for human tRNA locus chr19.trna13-ValCAC as an example

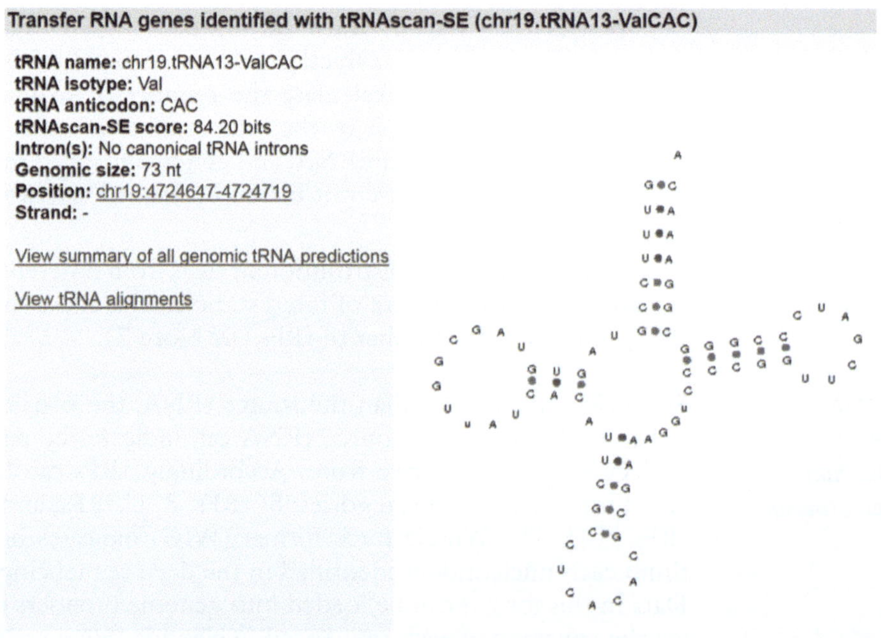

Fig. 3 Example of tRNAscan-SE information for human tRNA locus chr19.trna13-ValCAC

4. Further information can be extracted for each tRNA locus by clicking on the relevant tRNAscan-SE result at the bottom of the UCSC browser (shown by arrow in Fig. 2) to show a screen similar to Fig. 3.

3.4 Enrichment of tRFs by Biochemical Purification

Since sequencing of small RNA libraries can introduce a quantitative bias it is essential to verify the results of deep sequencing data by conventional quantitative biochemical techniques such as northern blotting or quantitative PCR (qPCR) (*see* **Note 11**). tRFs can be further analyzed biochemically following isolation from living cells. However, as tRFs can often represent a minor population of small RNAs, it is critical to separate them from more abundant RNA species such as miRNAs and also their full-length tRNA precursors. This section outlines a biochemical purification method for 5′ tRFs. *See* Fig. 4a for an outline of the workflow.

1. Harvest ten to fifteen 15 cm culture dishes using trypsin, wash cells three times with ice-cold PBS, and resuspend the cells in three pellet volumes of ice-cold hypotonic lysis buffer.
2. Incubate the cells for 7 min and then break cells with 10 strokes of a tight-fitting dounce homogenizer.
3. Check the efficiency of the lysis by adding Trypan Blue to a small aliquot of cells (mixing 1:1) and visualize them by light microscopy. The nuclei of lysed cells stain blue and clump together.
4. Centrifuge the homogenate (220 × *g* for 5 min at 4 °C) and recover the supernatant.
5. Mix the supernatant with 0.11 vol of Buffer B and centrifuge the samples in a swinging bucket rotor (60,000 × *g* for 1 h at 4 °C).
6. Dialyze the supernatant against an excess of Buffer D for 4–16 h. The resulting solution is cytosolic S100 extract (*see* **Note 12**). Make aliquots, flash freeze, and store them at −80 °C.
7. Set up the HPLC apparatus with a 1 mL ionic exchange column (Q-column) and add Ionic Exchange Buffers A (0 M NaCl) and B (1 M NaCl) to the appropriate reservoirs.
8. Set up an HPLC program to run the following steps:
   - Injection of the sample.
   - Washing of the column with Ionic Exchange Buffer A until the OD_{280} returns to baseline.
   - Stepping up of the NaCl concentration to 480 mM.
   - Gradient elution of fractions from the column with NaCl concentrations between 480 and 650 mM.
   - Washing and regeneration of the column with 1 M NaCl.
9. For downstream analysis, choose the salt elution that maximizes tRF recovery and minimizes miRNA contamination. This is shown in Fig. 4b, where fractions from this procedure have been probed for *miR-21* and the tRF *Gln19* by northern blot.
10. Collect best fraction from the anion exchange purification and (e.g., fraction 3, Fig 4b) load it onto a Superdex 200 gel filtration column. In a previous or subsequent run, molecular weight standards (Bio-Rad) should be run under identical conditions.
11. Elute the RNA fragments using Buffer E and collect all fractions. Combine the fractions corresponding to the 17–44 kDa standard range (*see* red box, Fig. 4c, which shows a northern blot of further enriched tRNA-derived fragments).

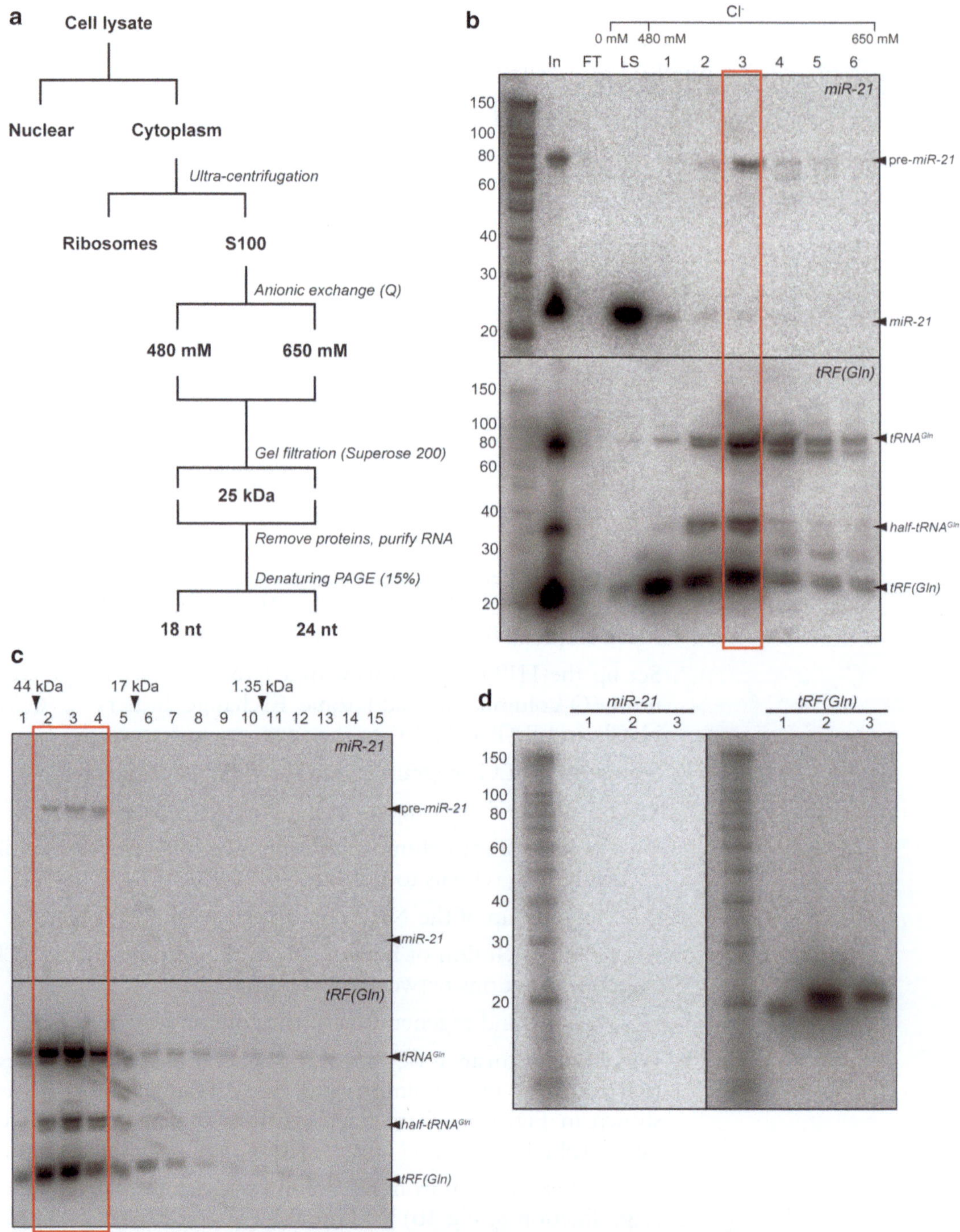

Fig. 4 Biochemical purification of tRFs. (**a**) Workflow of purification procedure. (**b**) *miR-21* (*upper panel*) and *tRF(Gln)* (*lower panel*) northern blots of Cytoplasmic S100 extract from Q column purification. Fractions were eluted with [NaCl] between 480 and 650 mM. *FT* flowthrough, *LS* low salt elution (0–480 mM). (**c**) *miR-21* (*upper panel*) and *tRF(Gln)* (*lower panel*) northern blots of material from fraction 3 of Q column purification that was size purified on a Superdex 200 gel filtration matrix. Fractions 2–4 were used for PAGE purification. (**d**) RNA of different size ranges from PAGE purification was subject to northern blotting using probes to *miR-21* (*left panel*) and *tRF(Glnn* (*right panel*). *Lane 1*: 18–20 nt. *Lane 2*: 20–22 nt. *Lane 3*: 22–24 nt. RNA from *lane 2* was used as purified tRFs. Size markers are shown in bp

12. Mix the samples with 1 vol of 2× PK buffer, add proteinase K to the combined fractions at a concentration of 200 μg/mL, and incubate at 65 °C for 30 min.
13. Extract the RNA by adding 1:1 phenol:chloroform:isoamyl alcohol (24:25:1) (pH 6.6), mix by vortexing, and centrifuge at 16,500 × *g* to separate the phases.
14. Remove the supernatant and precipitate the RNA by adding 3 vol of ethanol and 35 μg/mL glycogen; precipitate the RNA overnight at −80 °C.
15. Centrifuge the samples at 16,500 × *g* for 30 min at 4 °C.
16. Wash the pellet with 70 % (v/v) ethanol and resuspend the pellet in Formamide Loading Buffer.
17. Run the purified product from the gel filtration on a 15 % denaturing urea-PAGE gel next to the radiolabeled 10 bp ladder.
18. Excise the section of the gel corresponding to tRFs (*see* **Note 13**) and dice the fragments into 1 mm^3 cubes using a clean scalpel.
19. Incubate the gel pieces in 1 mL 1× PK buffer at 20 °C overnight with gentle agitation.
20. Recover the liquid with the RNA, and purify the RNA by phenol:chloroform:isoamyl alcohol (24:25:1) extraction (*see* **step 13**).
21. Precipitate the RNA by adding 3 vol of ethanol and 35 μg/mL glycogen and incubate overnight at −80 °C.
22. Centrifuge the samples at 16,500 × *g* for 30 min at 4 °C.
23. Wash the pellet with 70 % (v/v) ethanol and resuspend the RNA in an appropriate buffer.
24. This pellet now contains purified tRFs ready for further analysis by northern blotting and/or quantitative PCR (*see* **Note 11**). An example of the efficiency of this procedure is shown in Fig. 4d. Northern blot hybridization with probes to *miR-21* (the most abundant miRNA in HeLa cells) and the tRF(Gln) (*see* **Note 14**) was performed on tRF samples after PAGE purification. The sample shows very little contamination with miRNAs and/or full-length tRNAs.

4 Notes

1. This protocol is intended to serve as a guide to maximize the productivity and accuracy of tRF mapping. The necessary infrastructure, bioinformatic resources, and knowledge of mapping procedures are required to perform basic mapping of raw deep sequencing data.

2. The GtRNAdb has an inherent naming convention that consists of: [species]; [chromosome number]; [numbered tRNA gene on chromosome]; [tRNA isotype]; [anticodon] (e.g., Homo_sapiens_chr6.trna93-AlaAGC). It is recommended that each locus is named accordingly during the mapping procedure so as to retain this information for later reference.
3. This should be performed prior to any genomic mapping procedure, comparable to mapping miRNAs to miRBase.
4. miRNA hairpin sequences are available from miRBase (*see* Subheading 2.3).
5. The gtRNAdb website also contains additional information, such as isotype distribution, locus information, the presence of introns, and predicted secondary structures. These can be compiled using the unique identifiers in a species-specific approach.
6. The tRNAscan-SE score is equivalent to a Cove score which is used to predict the likelihood that a sequence is tRNA-like [14].
7. tRNA loci with no identified reads will typically not appear in deep sequencing data. This is important if showing the overall distribution of tRF isotypes.
8. This is specifically performed by clicking "Add custom tracks." Ensure that the genome and assembly settings match those used for mapping (e.g., Human, hg19/GRch37).
9. If the track does not appear, ensure that the viewing type for the custom WIG track is set to "Full."
10. To extract the visual information on screen, images can be downloaded in PDF and/or postscript formats by accessing the website header menu "View" and selecting "PDF/PS."
11. We strongly recommend using the northern blotting protocol developed by Pall and Hamilton, 2008 [1](Fig. 5a). For quantitative PCR detection of tRFs in total RNA, we suggest Custom TaqMan small RNA assays (Invitrogen) developed by Chen et al., 2005 [15]. An example of a real-time TaqMan PCR analysis demonstrating the power of this assay to discern short tRF fragments is shown in Fig. 5b.
12. This method was optimized using S100 extract from the HeLa cell line.
13. tRFs have been shown to possess a variety of sizes. When choosing the size to excise, customize the region to your requirements.
14. tRF(Gln) is a previously well-described small tRF [5].

Acknowledgements

This work was supported by the ARC DP 130103027. G.H. is an ARC Future Fellow.

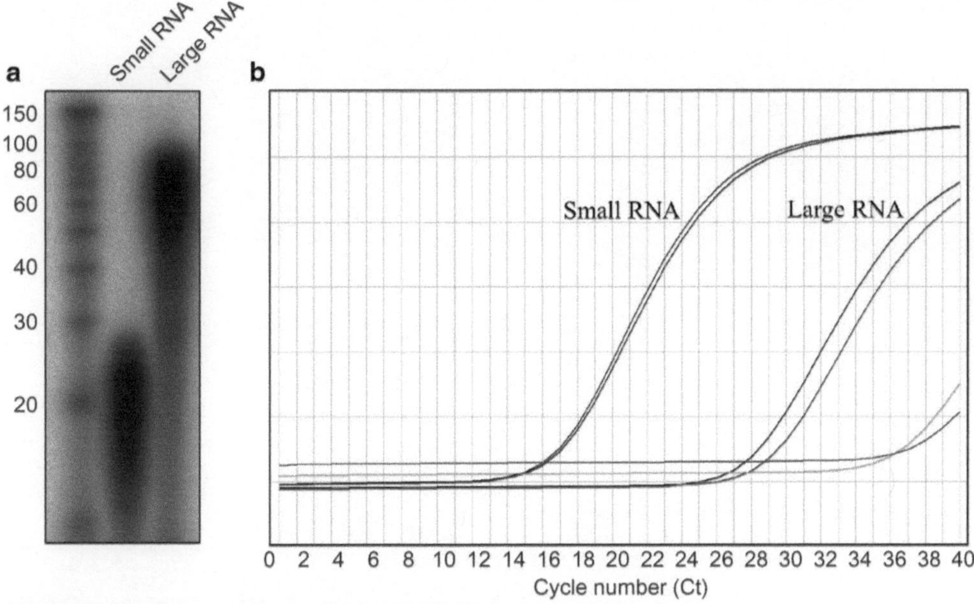

Fig. 5 Example of TaqMan quantitative PCR on enriched RNA samples. (**a**) HeLa RNA was separated into a large (tRNA-containing) and small (tRF-containing) fractions by denaturing PAGE purification. The fractions were radiolabeled using the PNK exchange reaction and visualized via autoradiography. (**b**) qPCR signal during amplification with primers to detect tRF(Gln). The Ct value for detection in the small RNA sample is 17.3; in the large sample it is 28.8

References

1. Pall GS, Hamilton AJ (2008) Improved northern blot method for enhanced detection of small RNA. Nat Protoc 3(6):1077–1084
2. Cole C et al (2009) Filtering of deep sequencing data reveals the existence of abundant Dicer-dependent small RNAs derived from tRNAs. RNA 15(12):2147–2160
3. Lee YS et al (2009) A novel class of small RNAs: tRNA-derived RNA fragments (tRFs). Genes Dev 23(22):2639–2649
4. Ivanov P et al (2011) Angiogenin-induced tRNA fragments inhibit translation initiation. Mol Cell 43(4):613–623
5. Sobala A, Hutvagner G (2013) Small RNAs derived from the 5′ end of tRNA can inhibit protein translation in human cells. RNA Biol 10(4):553–563
6. Gebetsberger J et al (2012) tRNA-derived fragments target the ribosome and function as regulatory non-coding RNA in Haloferax volcanii. Archaea 2012:260909
7. Haussecker D et al (2010) Human tRNA-derived small RNAs in the global regulation of RNA silencing. RNA 16(4):673–695
8. Wang Q et al (2013) Identification and functional characterization of tRNA-derived RNA fragments (tRFs) in respiratory syncytial virus infection. Mol Ther 21(2):368–379
9. Kent WJ et al (2002) The human genome browser at UCSC. Genome Res 12(6): 996–1006
10. Chan PP, Lowe TM (2009) GtRNAdb: a database of transfer RNA genes detected in genomic sequence. Nucleic Acids Res 37(Database issue): D93–D97
11. Schattner P, Brooks AN, Lowe TM (2005) The tRNAscan-SE, snoscan and snoGPS web servers for the detection of tRNAs and snoRNAs. Nucleic Acids Res 33(Web Server issue): W686–W689
12. Kozomara A, Griffiths-Jones S (2011) miRBase: integrating microRNA annotation and deep-sequencing data. Nucleic Acids Res 39(Database issue):D152–D157
13. Sobala A, Hutvagner G (2011) Transfer RNA-derived fragments: origins, processing, and functions. Wiley Interdiscip Rev RNA 2(6):853–862
14. Eddy SR, Durbin R (1994) RNA sequence analysis using covariance models. Nucleic Acids Res 22(11):2079–2088
15. Chen C et al (2005) Real-time quantification of microRNAs by stem-loop RT-PCR. Nucleic Acids Res 33(20):e179

Chapter 15

Differential DNA Methylation Patterns in Endo-siRNAs Mediated Silencing of LINE-1 Retrotransposons

Long Chen, Jane E Dahlstrom, and Danny Rangasamy

Abstract

Analyzing differences in DNA methylation is a powerful tool for assessing the effect of endo-siRNAs expression in the human genome. Here, we present a simple genome-wide DNA methylation assay that allows for a precise quantitative analysis of differences in the promoter of human long interspersed nuclear element 1 (LINE-1 or L1) retrotransposons in response to endogenous and exogenous expression of endo-siRNAs. Using the DNA bisulfite modification sequencing, we have optimized the method to detect small changes in heterogeneously methylated L1 repeats at multiple regions across the genome. We also provide guidance for analysis of primary bisulfite sequencing data and interpretation of the methylation status using the Web-based bisulfite sequencing DNA methylation (BISMA) analysis. This refined and reproducible assay can be performed even using a small amount of genomic DNA and is suitable for the analysis of clinical tissue samples.

Key words Endo-siRNAs, LINE-1, Breast cancer cells, DNA methylation, DNA sequencing, PCR, Bisulfite data analysis

1 Introduction

During tumorigenesis, cells undergo both genome-wide DNA hypomethylation and local hypermethylation of gene promoters, resulting in transcriptionally inactive genes becoming active, and genes that are normally active rendered inactive due to the changes of DNA methylation [1]. Long interspersed nuclear element 1 (LINE-1 or L1) is a repetitive DNA element and is distributed throughout the genome. About 20 % of the human genome is made up of L1 retrotransposons, which are capable of reverse transcribing their own mRNAs and inserting DNA copies into new places within the genome [2]. In the majority of normal cells, L1 retrotransposons are rarely expressed and are heavily methylated at CpG islands of their promoters, rendering them inactive [3, 4].

Hypomethylation of L1 sequences occurs at a very early stage of malignant transformation, resulting in aberrant expression of L1

Andreas Werner (ed.), *Animal Endo-siRNAs: Methods and Protocols*, Methods in Molecular Biology, vol. 1173, DOI 10.1007/978-1-4939-0931-5_15, © Springer Science+Business Media New York 2014

retrotransposons [5, 6]. In humans, L1 methylation has been shown to be highly variable at its promoter [7] and hypomethylation is strongly associated with poor prognosis of patients [5]. This genome-wide loss of L1 DNA methylation is often regarded as a common epigenetic event in tumorigenesis. Indeed, the majority of DNA methylation studies have used L1 as a surrogate DNA marker for estimating global methylation status of cancer genomes [8].

At present, at least 100 copies of L1 repeats have been identified as functional elements [9], retaining their ability to alter the genome by disrupting genes, altering genes splicing, increasing the frequency of recombination, and negatively affecting the stability and integrity of the genome due to their ability to create double-stranded DNA breaks during the process of retrotransposition [10]. It is believed that L1 expression is held in check by a variety of genome defense mechanisms including DNA methylation, RNAi-mediated silencing pathways, and repressive heterochromatin structures.

Several studies performed in mice and *Drosophila* [11, 12] have shown that naturally occurring endogenous siRNAs (endo-siRNAs) can repress L1 expression through an RNAi pathway, acting as a constant genome defense system. Endo-siRNAs are small noncoding RNAs that control target gene expression through their secondary structure and by recruiting chromatin-targeted RNAi silencing components to form transcriptionally inactive heterochromatin structures [13]. Using high-throughput deep sequencing, we have recently identified a subset of human L1-specific endo-siRNAs that are differentially expressed between normal breast cells and breast cancer cells [4]. When targeted to the L1 promoter, endo-siRNAs load onto the RNAi machinery protein Argonaute and direct the silencing of L1 elements. Notably, the overexpression of the identified endo-siRNAs in breast cancer cells resulted in marked silencing of L1 expression through hypermethylation of the L1 promoters. Because L1 methylation is highly variable in human cells, methods used need to readily detect small changes in DNA methylation of L1 repeats at multiple regions across the genome. While several methods exist for analyzing L1 methylation, bisulfite sequencing is of high interest because of its ability to detect and compare DNA methylation density between two samples at a single CpG dinucleotide level.

Sequencing of bisulfite converted DNA is still the preferred method in the field of cancer research and epigenetics for analyzing DNA methylation. Genomic DNA isolated from cells or tissues is treated with sodium bisulfite, which can selectively deaminate unmethylated cytosine to uracil but not methylated cytosine. This leads to a primary sequence change in the DNA that allows one to distinguish unmethylated cytosine from methylated cytosine. The sequence differences between a methylated and unmethylated cytosine can be analyzed by multiple methods such as PCR

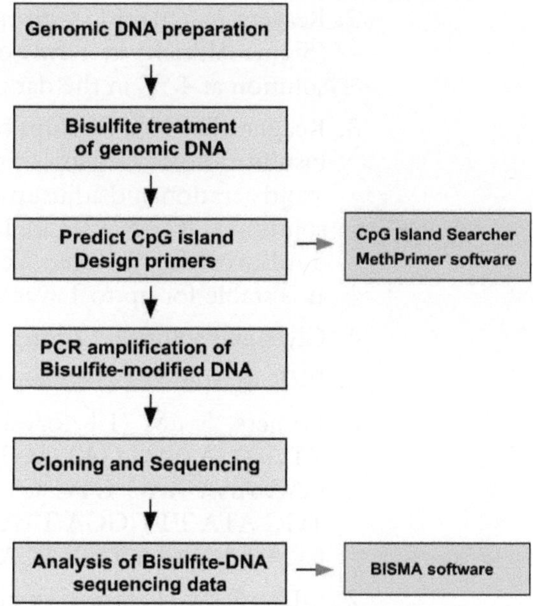

Fig. 1 Flowchart of bisulfite DNA sequencing analysis

amplification, direct sequencing, combined bisulfite restriction analysis (COBRA), methylation specific PCR (MSP), and pyrosequencing. The most popular method is PCR amplification combined with direct cloning and sequencing of a region of interest.

Here, we present a simple and relatively easy method of bisulfite DNA sequencing to quantitatively analyze differences in L1 methylation in breast cancer cells treated with or without overexpression of L1-specific endo-siRNAs. This modified assay can be performed with starting DNA amounts of as little as 50 ng up to 1 μg of genomic DNA. Figure 1 shows the outline of the method which describes the following steps: bisulfite treatment of DNA; primer design and PCR amplification; and methodology for sequence analysis and data processing. This step-by-step protocol offers a quick and cost-effective approach for detecting minute changes in the CpG methylation status of L1 retrotransposons among a large pool of heterogeneously methylated L1 repetitive DNA templates.

2 Materials

Prepare all solutions using sterile water and molecular biology grade reagents.

1. 3 M NaOH: 100 mg NaOH in 830 μL water. Mix and incubate in a water bath at 55 °C for 2–3 min for complete dissolution. Cool down to room temperature before use. This solution should be prepared freshly prior to each use.

2. Reagent A—20 mM Hydroquinone: Mix 2.2 mg of hydroquinone (Sigma-Aldrich) in 1 mL of water by vortexing and keep the solution at 4 °C in the dark by covering with aluminum foil.
3. Reagent B—40 % Sodium bisulfite: Dissolve 2.025 g of sodium bisulfite (Acros Organics) in 4 mL water by gentle inversion to avoid aeration and adjust pH to 5 with 10 M NaOH (this stock solution is 4 g NaOH in 10 mL water) and then make up to 5 mL with sterile water. Keep the solution at 4 °C in the dark. It is stable for up to 1 week.
4. Glycogen solution (20 mg/mL, Sigma-Aldrich).
5. PCR Master Mix (Promega).
6. Primers, 5 μM (L1 forward: 5′-TTA TAA ATT ATG TTT TTG TGA ATG GAT AGT-3′; L1 reverse: 5′-TTA CAA AAA CAA AAA AAC CTC CT-3′; α-SAT forward: 5′-TGT AAG TGG ATA TTT GGA TTA TTG G-3′; α-SAT reverse: 5′-TTT CCA AAA AAA TCT TCA AAA AAT -3′).
7. QIAquick Gel extraction kit (Qiagen).
8. QIAprep Spin Miniprep kit (Qiagen).
9. Rapid DNA ligation System (Promega).
10. pGEM-T Vector (Promega) or TOPO Cloning kit (Invitrogen).

3 Methods

3.1 DNA Preparation

1. Isolate genomic DNA from cell lines or tissue samples using any of the standard protocols. DNA should be of good quality and be free from any impurities. Extraction using the Arcturus PicoPure DNA kit is highly recommended for small samples obtaining from paraffin-embedded, formalin-fixed tissues retrieved using Laser Capture Microdissection. If this is not possible, the DNA should be treated with standard phenol–chloroform extraction before use.
2. Using a NanoDrop UV–Vis spectrometer or any other spectrophotometer of your choice, measure DNA concentrations at 260 nm and purity using the 260/280 ratios (*see* **Note 1**). At least 2 μL of DNA sample is required to accurately measure the genomic DNA concentration without dilution.
3. Transfer the DNA into a 1.5 mL microcentrifuge tube. If a thermal cycler is available, use 0.5 mL thin-walled tubes with individual lids. Starting DNA can vary from 50 ng to 1 μg diluted in 20 μL of water or TE buffer (10 mM Tris–HCl, pH 7.5, 1 mM EDTA). Store the DNA at 4 °C until you are ready for next step.

3.2 Bisulfite Preparation

1. Add 2.2 μL of 3 M NaOH solution to the 20 μL of DNA solution. Mix well by gentle pipetting and incubate at 37 °C for 20 min in a thermal cycler or water bath.
2. In the meantime prepare fresh bisulfite solution by mixing and vortexing 7.5 μL of Reagent A and 212.5 μL of Reagent B for each reaction (*see* **Note 2**). The combined solutions should be kept at 4 °C in the dark or used immediately. Alternatively, several commercial kits are available for this step. The MethylEasy Xceed kit (Human Genetic Signatures, Australia) works well in our hands particularly when using small amounts of genomic DNA. The reagents in this commercial kit need to be premixed before use (*see* **Note 3**).
3. Add 220 μL of combined reagents A and B to each DNA sample. Gently mix by pipetting up and down and overlay with 100 μL of mineral oil (optional) to prevent evaporation (*see* **Note 4**). Incubate the reaction at 55 °C for 4–16 h in a water bath or a thermal cycler (*see* **Note 5**). If possible, cover the tube with aluminum foil to protect from light.
4. Cool the sample briefly on ice and centrifuge tubes for 5–10 s to bring down any condensate. Remove the sample from underneath the mineral oil using a pipette tip and transfer the sample to a new centrifuge tube. Add 1 μL glycogen solution (corresponding to 20 μg glycogen) (*see* **Note 6**) and mix by pipetting.
5. Denature the DNA samples by adding 24.7 μL of 3 M NaOH (to a final concentration of 0.3 M) and incubate at 37 °C for 15 min.
6. Purify the DNA using QIAquick spin columns as described by the manufacturer's instructions or any other column-based method of your choice. Elute the DNA with 50 μL of warm water or TE buffer and store it at −20 °C until you perform the PCR amplification.
7. If the starting DNA is less than 100 ng, column purification of the DNA can be omitted; instead, the DNA can be precipitated. Add 1 mL of 100 % isopropanol to the DNA samples and then vortex for 5 s. Incubate at 4 °C for at least 60 min and centrifuge the tubes at maximum speed (14,000 to 16,000 × *g*) for 20 min. Carefully discard the supernatant (*see* **Note 7**).
8. Wash the pellet with 70 % ethanol, centrifuge for 15 min at maximum speed, and discard the supernatant. Allow the pellet to air-dry for 15–30 min at room temperature. Resuspend the DNA pellet in 25 μL TE buffer and store it at −20 °C until you are ready for PCR amplification.

3.3 Amplification of Bisulfite-Modified DNA

1. Design the primers based on a consensus sequences of the L1 promoters that allow for the amplification of a pool of the L1 CpG islands (*see* **Note 8**). The 560 bp amplicon includes a total of 31 CpG sites between the primers, representing global methylation status of L1 DNA. It is recommended to add control PCR reactions to amplify the centromeric alpha satellite (α-SAT) sequences to assess the integrity and sensitivity of the bisulfite-modified DNA.
2. Set up the PCR reactions using 0.2 mL thin-walled PCR tubes containing the following reactions: 12.5 μL of PCR Master Mix (Promega), 1 μL of L1 forward primer (0.5 μM), 1 μL of L1 reverse primer (0.5 μM), 1–2 μL of bisulfite-converted DNA (~10 ng), and nuclease-free water to 25 μL. In a separate tube, add 1 μL each of α-SAT forward and reverse primers instead of L1 primers (positive control). In another tube, add 1–2 μL untreated DNA template (i.e., without bisulfite conversion) as a negative control. Keep all the tubes on ice and spin briefly to collect the material at the bottom of the PCR tubes.
3. Carry out the PCR reaction using the following conditions: 95 °C for 4 min, followed by 35 cycles of 95 °C for 45 s, 51 °C for 90 s, 72 °C for 120 s and a final step of 72 °C for 8 min (*see* **Note 9**).
4. Load the PCR products on a 1.5 % agarose gel. The expected size of bisulfite PCR products is 223 bp for α-SAT and 560 bp for L1 DNA. To determine where the expression of the L1-specific endo-siRNAs results in hypermethylation of the L1 promoter, the methylation levels of L1 promoters can be measured in the endo-siRNAs and controls-transfected normal breast HMECs and MCF10A cells, as well as breast cancer T47D cells (Fig. 2). As positive controls, α-Satellite DNA is amplified with primers that should only yield a PCR products with bisulfite-modified DNA but not with untreated DNA.
5. The PCR products can be purified using the QIAquick Gel extraction kit. Elute the DNA in 30 μL prewarmed TE buffer. Store the DNA at −20 °C until ready for cloning.

3.4 Cloning and Sequencing

1. Vortex the 2× Rapid Ligation buffer vigorously before use. Set up the ligation reaction in a 0.5 mL tube containing: 2.5 μL of 2× Rapid Ligation buffer, 0.5 μL of pGEM-T vector (25 ng), 1.5 μL of PCR product, and 0.5 μL of T4 DNA Ligase (3 U/μL). Mix and incubate the reaction at room temperature for 1–3 h.
2. Transfer 40 μL of electrocompetent DH5α or TOP10 cells to 1 μL of the ligation reaction in a prechilled cuvette and electroporate the cells with standard setting of 2.5 KV, 25 μF capacitance, and 200 Ω resistance using Bio-Rad electroporator (*see* **Note 10**). Alternatively, 1 μL of ligation product can be used

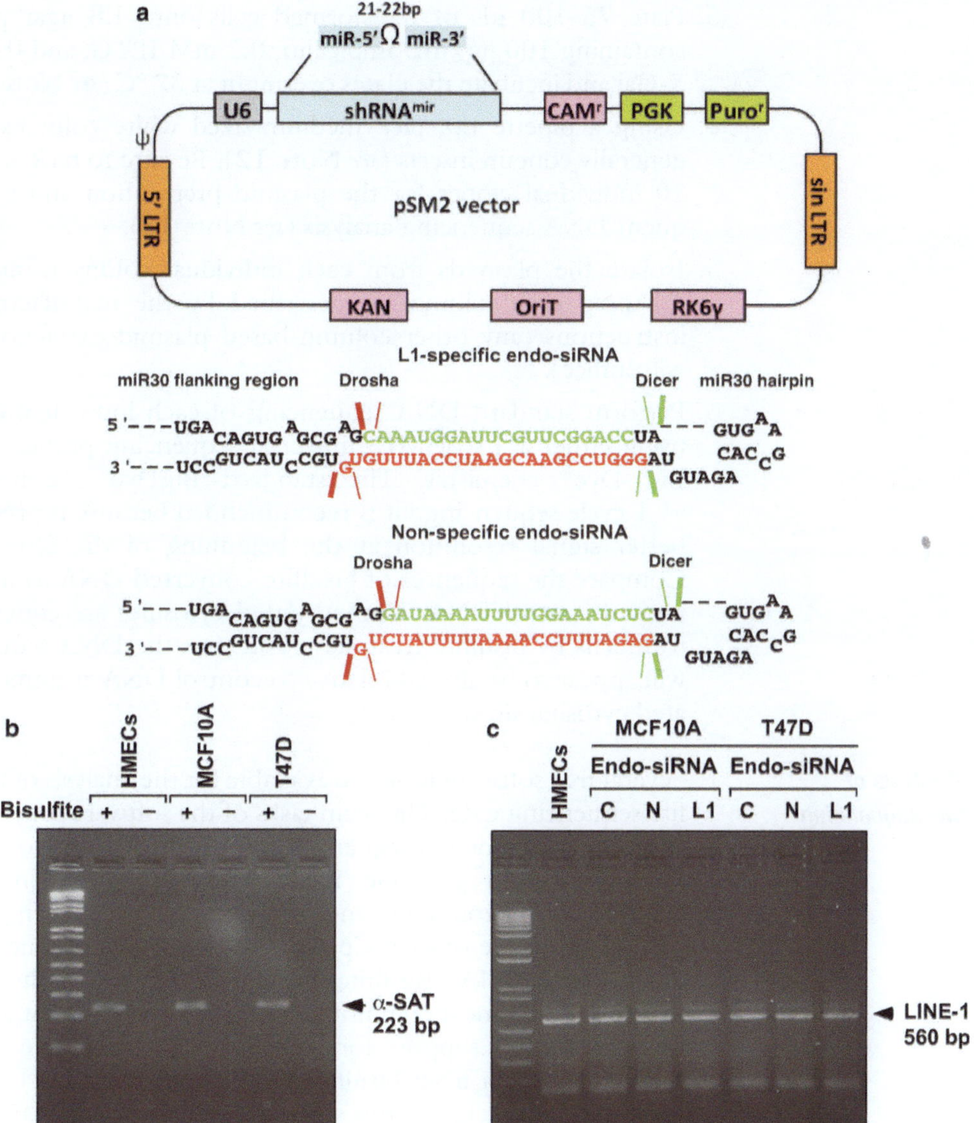

Fig. 2 PCR amplification of bisulfite-modified L1 DNA (**a**) A schematic representation of the pSM2 vector used for the expression of endo-siRNAs is shown in the *top panel*. The synthetic L1-specific endo-siRNA and Nonspecific control-endo-siRNAs used in this study are shown at the *bottom*. (**b**) Amplification of alpha satellite DNA with methylated (shown as + symbol) and unmethylated (shown as – symbol) genomic DNA of breast cancer cells. (**c**) Representative gel picture showing the PCR amplification of the CpG island in the L1 promoter. Bisulfite-modified HMEC tissue was used as a positive control. The parental cells and cells transfected with control endo-siRNA or L1-specific endo-siRNA were performed in parallel. The symbol C represents the untransfected parental cells, while the symbols N and L1 represent the cells transfected with the control-endo-siRNA and the L1-specific endo-siRNA transfected cells, respectively

to transform chemically competent *E.coli* cells by heat-shock at 42 °C for 45–50 s in a water bath. Add 960 μL SOC medium and incubate the cells at 37 °C for 45–60 min with 150 rpm shaking.

3. Plate 75–100 μL of transformed cells onto LB agar plates containing 100 μg/mL ampicillin, 0.2 mM IPTG, and 0.04 % X-Gal and incubate the plates overnight at 37 °C (*see* **Note 11**).
4. Using a pipette tip, pick medium-sized white colonies that generally contain inserts (*see* **Note 12**). Be sure to pick at least 20 individual clones for the plasmid preparation and subsequent DNA sequencing analysis (*see* **Note 13**).
5. Isolate the plasmids from each individual colony using the QIAprep spin columns as described by the manufacturer's instructions (any other column-based plasmid extraction kit will suffice).
6. Perform standard DNA sequencing of each individual clone using either a T7 or M13 universal sequencing primers with Big Dye® chemistry. The standard BigDye® Terminator v1.1 cycle sequencing kit is recommended because it provides better signal resolution at the beginning of the sequence. Compare the sequences of bisulfite-converted DNA to mock-treated control DNA. Unmethylated cytosines are converted to uracils by bisulfite treatment, and thus the DNA sequence will appear to be altered relative to control DNA at unmethylated cytosine sites.

3.5 Analysis of Bisulfite-Sequencing Data

1. Several free software tools are available for the analysis of bisulfite sequencing data. The main tasks of the software tool are to align the experimental sequences of individual clones to the in silico converted genomic DNA reference, measuring the bisulfite-conversion rate, and analyzing and visualizing the methylation state of each CpG site in the bisulfite sequencing data. The BISMA (Bisulfite Sequencing DNA Methylation Analysis) is the most commonly used software tool that allows for start-to-end support for processing the repetitive DNA methylation data from bisulfite DNA sequencing [14]. Here we focus on how to use this software tool for quantification of the L1 promoter methylation status in the presence or absence of the endo-siRNAs expression in normal MCF10A cells and breast cancer T47D cells.
2. BISMA is an online tool for the analysis of bisulfite sequencing of repetitive DNA elements. The step-by step guided tour for using this tool is available for occasional users in http://biochem.jacobs-university.de/BDPC/BISMA/. This software tool has a highly automated data processing algorithm and calculates the average methylation at each CpG sites and displays methylation pattern of multiple individual clones. The first step in this tool is to copy and paste the unmodified sequence of the L1 promoter region and then import all the

experimental sequences that were bisulfite-modified and sequenced. The second step is setting the default filtering thresholds at 90 % conversion rate (*see* **Note 14**). The program will automatically detect the proper DNA direction; remove sequences that do not align with the reference sequence, and then calculate a clustalW alignment of the sequence and the percentage of methylation conversion rates. The final output will not only display a comparison of cytosine patterns but also sort the sequence based on methylation levels for statistical analysis as well as present the data graphically. The advantage of using the BISMA tool for repetitive L1 methylation analysis is the pairwise comparison of the cytosine patterns of all sequences and ignoring unresolved N-sites, deletions, and mutations of repetitive sequences that were bisulfite-modified and sequenced. Additionally, BISMA also visualizes the distribution of CpG sites among individual sequences. Figure 3 illustrates the analysis of DNA methylation of L1 promoters in breast cancer cells in the presence of endo-siRNAs as compared to cells with empty vector.

4 Notes

1. Absorbance of the sample at 260 nm in a NanoDrop represents 10 times the absorbance actually measured using the standard 1 mm path length. NanoDrop shows a sample concentration in ng/μL based on absorbance at 260 nm. The purity of DNA should be within a ratio of 1.8–2.0 at 260/280 nm.
2. Calculate the number of DNA samples that need to be treated and include an aliquot to allow for pipetting errors.
3. The combined solution should be kept in the dark. Both hydroquinone and sodium bisulfite solutions are light sensitive and thus avoid unnecessary exposure to light by keeping in the dark and covering with aluminum foil. Although the combined solution is stable for up to 1 week at 4 °C in the dark, it is recommended that the solution should be fresh.
4. Overlaying with mineral oil can be omitted by using a thermal cycler and thin-walled tubes with individual lids.
5. Bisulfite conversion can be carried out in as little as 4 h; however, the longer incubation time of 16 h or overnight results in full methylation conversion.
6. Addition of glycogen is strongly recommended if starting DNA is less than 100 ng. Alternatively, DNA can be precipitated using 0.5 μg of tRNA or salmon sperm carrier DNA (Sigma-Aldrich, D9156).

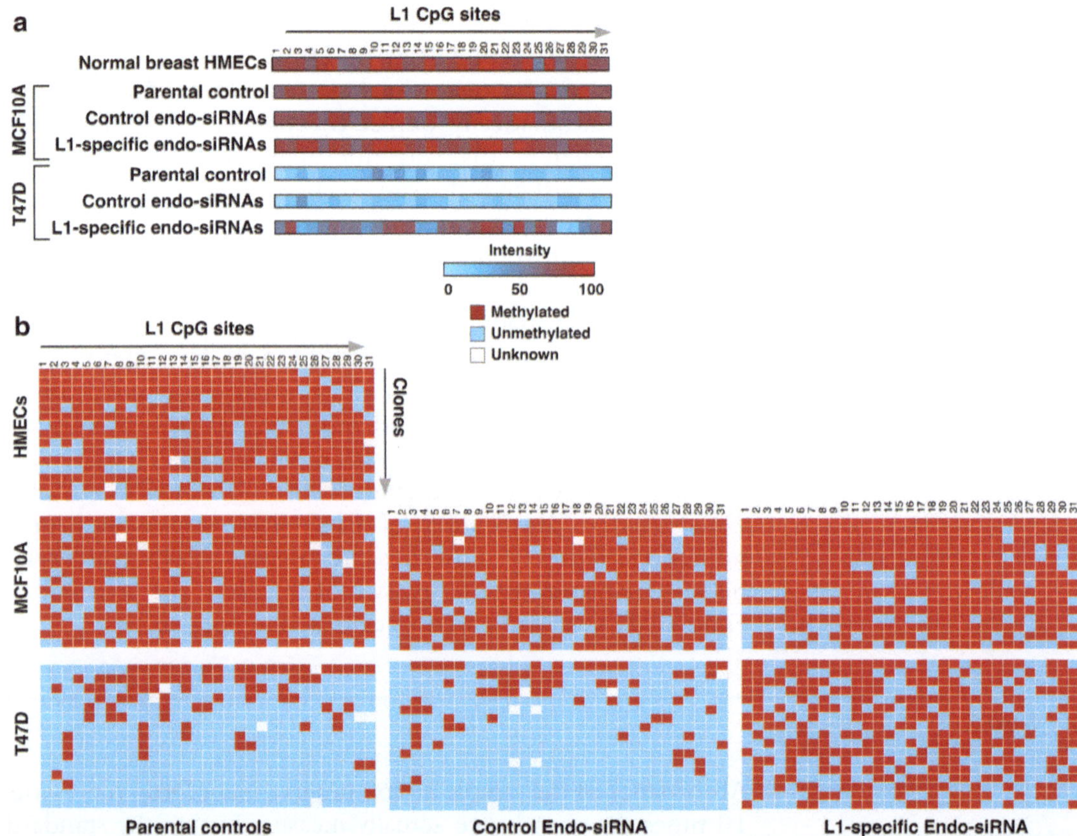

Fig. 3 Analysis of L1 DNA methylation using the BISMA software tool. (**a**) Graphical representation of the BISMA output showing the methylation status of L1 promoters in the presence or absence of the endo-siRNA overexpression. Bisulfite-modified normal breast HMECs and MCF10A cells were used as a positive control. The parental cells and cells transfected with control endo-siRNA or L1-specific endo-siRNA were performed in parallel. Methylated CpG sites are shown in *red*, while unmethylated and undetermined CpG sites are shown in *blue* and *white* colors, respectively. (**b**) Comparison of each individual L1 CpG sites derived from the multiple individual clones. Each column represents the DNA methylation status of each CpG sites analyzed and each row corresponds to single clones analyzed which were subcloned and sequenced. The *red square* indicates high levels of CpG methylation while *blue squares* are unmethylated sites

7. Do not incubate at –20 °C as isopropanol efficiently precipitates only DNA at 4 °C or room temperature. The DNA pellet may not be visible after isopropanol precipitation, and therefore when removing the supernatant, keep the pipette tip away from the expected location of the DNA pellet.

8. One of the critical issues for successful bisulphite analysis of DNA methylation is careful designing of high-quality primers that specifically amplify CpG islands within the L1 promoters. Primers should be designed at the target regions with at least 3–4 cytosines (which will be converted to T's) but CpG depleted regions. This not only enhances the "specificity" to

accurately amplify bisulfite-modified DNA sequences but also eliminates not fully converted DNA sequences. Alternatively, one can use the free software available from Applied Biosystems, Methyl Primer Express® Software v1.0, which enables to design PCR primers for methylation mapping studies by simply cutting and pasting the DNA sequences of interest. This software searches for CpG islands by simulating in silico bisulfite modification of DNA sequences, and highlighting suitable target regions and primers for methylation analysis.

9. Nested PCR or second round PCR amplification is not recommended for the amplification of bisulfite-modified L1 sequences because of the high copies of L1 repeats in the human genome and the possibility of cross contamination and amplification bias. Amplifying bisulfite converted L1 DNA in our hands does not seem to be a problem. However, it is important to use freshly converted DNA (no more than a week old) in the PCR reactions to obtain clean results.

10. Both DH5α and TOP10 cells work equally well in our hands for cloning the bisulfite-converted DNA. We have also tested other bacterial strains such as SURE and chemically competent JM109 cells, and have not found any difference with respect to the *E.coli* strain used. However, we noticed that substituting SOC medium with LB broth decreases the number of colonies on LB plates.

11. Spread the cells uniformly on plates to get enough individual clones. In our hands, we typically obtain approximately 30–40 colonies per plates. Given that L1 is a repetitive element, the orientation of some of the cloned fragments often influences the number of bacterial colonies grown on the LB plates. If a higher number of colonies are required, the transformed cells can be centrifuged at $500 \times g$ for 10 min and resuspended in 200 μL of SOC medium. Plate 100 μL of cells on two LB plates.

12. Colonies containing no inserts often grow slowly compared to cells containing inserts. The blue colonies represent no inserts because of β–galactosidase activity. Select white colonies of medium size (approximately 1 mm in diameter).

13. Efficiency of bisulfite modification may vary between the clones. Mutation often occurs in the L1 promoters, and thus the assay cannot distinguish the origin of TpG from either mutation of 5 mC or conversion of an unmethylated cytosine. Selecting more individual colonies will aid in the reproducibility of results.

14. It is important that all sequences with an unacceptably low bisulfite conversion rate (below a 90 % cutoff value) or high sequencing errors are removed for high-quality data processing and analysis.

Acknowledgments

This work was supported by a grant from the Canberra Hospital Private Practice Fund and the ACT Health and Medical Research Support Program.

References

1. Ehrlich M (2002) DNA methylation in cancer: too much, but also too little. Oncogene 21: 5400–5413
2. Beck CR, Collier P, Macfarlane C, Malig M, Kidd JM, Eichler EE, Badge RM, Moran JV (2010) LINE-1 retrotransposition activity in human genomes. Cell 141:1159–1170
3. Roman-Gomez J, Jimenez-Velasco A, Agirre X, Cervantes F, Sanchez J, Garate L, Barrios M, Castillejo JA, Navarro G, Colomer D, Prosper F, Heiniger A, Torres A (2005) Promoter hypomethylation of the LINE-1 retrotransposable elements activates sense/antisense transcription and marks the progression of chronic myeloid leukemia. Oncogene 24:7213–7223
4. Chen L, Dahlstrom JE, Lee SH, Rangasamy D (2012) Naturally occurring endo-siRNA silences LINE-1 retrotransposons in human cells through DNA methylation. Epigenetics 7:758–771
5. van Hoesel AQ, van de Velde CJ, Kuppen PJ, Liefers GJ, Putter H, Sato Y, Elashoff DA, Turner RR, Shamonki JM, de Kruijf EM, van Nes JG, Giuliano AE, Hoon DS (2012) Hypomethylation of LINE-1 in primary tumor has poor prognosis in young breast cancer patients: a retrospective cohort study. Breast Cancer Res Treat 134:1103–1114
6. Chen L, Dahlstrom JE, Chandra A, Board P, Rangasamy D (2012) Prognostic value of LINE-1 retrotransposon expression and its subcellular localization in breast cancer. Breast Cancer Res Treat 136:129–142
7. Estecio MR, Gharibyan V, Shen L, Ibrahim AE, Doshi K, He R, Jelinek J, Yang AS, Yan PS, Huang TH, Tajara EH, Issa JP (2007) LINE-1 hypomethylation in cancer is highly variable and inversely correlated with microsatellite instability. PLoS One 2:e399
8. Yang AS, Estecio MR, Doshi K, Kondo Y, Tajara EH, Issa JP (2004) A simple method for estimating global DNA methylation using bisulfite PCR of repetitive DNA elements. Nucleic Acids Res 32:e38
9. Brouha B, Schustak J, Badge RM, Lutz-Prigge S, Farley AH, Moran JV, Kazazian HH Jr (2003) Hot L1s account for the bulk of retrotransposition in the human population. Proc Natl Acad Sci U S A 100:5280–5285
10. Cordaux R, Batzer MA (2009) The impact of retrotransposons on human genome evolution. Nat Rev 10:691–703
11. Watanabe T, Totoki Y, Toyoda A, Kaneda M, Kuramochi-Miyagawa S, Obata Y, Chiba H, Kohara Y, Kono T, Nakano T, Surani MA, Sakaki Y, Sasaki H (2008) Endogenous siRNAs from naturally formed dsRNAs regulate transcripts in mouse oocytes. Nature 453: 539–543
12. Okamura K, Chung WJ, Ruby JG, Guo H, Bartel DP, Lai EC (2008) The Drosophila hairpin RNA pathway generates endogenous short interfering RNAs. Nature 453:803–806
13. Moazed D (2009) Small RNAs in transcriptional gene silencing and genome defence. Nature 457:413–420
14. Rohde C, Zhang Y, Reinhardt R, Jeltsch A (2010) BISMA–fast and accurate bisulfite sequencing data analysis of individual clones from unique and repetitive sequences. BMC Bioinformatics 11:230

Index

Andreas Werner (ed.), *Animal Endo-siRNAs: Methods and Protocols*, Methods in Molecular Biology, vol. 1173, DOI 10.1007/978-1-4939-0931-5, © Springer Science+Business Media New York 2014

P

Q

R

S

T

V

X

MIX
Papier aus verantwortungsvollen Quellen
Paper from responsible sources
FSC® C105338

If you have any concerns about our products,
you can contact us on
ProductSafety@springernature.com

In case Publisher is established outside the EU,
the EU authorized representative is:
Springer Nature Customer Service Center GmbH
Europaplatz 3, 69115 Heidelberg, Germany

Printed by Libri Plureos GmbH
in Hamburg, Germany